PRACTICAL SDR

PRACTICAL SDR

Getting Started with Software-Defined Radio

by David Clark and Paul Clark

no starch press®

San Francisco

Printed in the United States of America

First printing

29 28 27 26 25   1 2 3 4 5

ISBN-13: 978-1-7185-0254-3 (print)
ISBN-13: 978-1-7185-0255-0 (ebook)

 Published by No Starch Press®, Inc.
245 8th Street, San Francisco, CA 94103
phone: +1.415.863.9900
www.nostarch.com; info@nostarch.com

Publisher: William Pollock
Managing Editor: Jill Franklin
Production Manager: Sabrina Plomitallo-González
Production Editor: Sydney Cromwell
Developmental Editor: Nathan Heidelberger
Cover Illustrator: Joshua Kemble
Interior Design: Octopod Studios
Technical Reviewer: Josh Mormon
Copyeditor: Lisa McCoy
Proofreader: Daniel Wolff
Indexer: BIM Creatives, LLC

Library of Congress Control Number: 2024041499

For customer service inquiries, please contact info@nostarch.com. For information on distribution, bulk sales, corporate sales, or translations: sales@nostarch.com. For permission to translate this work: rights@nostarch.com. To report counterfeit copies or piracy: counterfeit@nostarch.com.

[S]

To MQ and LVP
Y and S and K.

To the infinitely supportive Jennifer, and to Jessica, who
brightens every day.

And to our parents, who invested in love and education.

About the Authors

David Clark is an engineer interested in how things work, and he enjoys sharing what he learns. He has been working with radio technology since the late 1980s and was using software-defined radios (SDRs) before they were cool.

Paul Clark is the owner and chief engineer at Factoria Labs, an organization dedicated to the propagation of SDR. He has experience ranging from chip design to firmware development to radio frequency reverse engineering. He teaches classes and workshops on SDR in the United States and abroad.

About the Technical Reviewer

Josh Mormon is a senior research scientist in the wireless communications industry, where he works primarily on SDR applications. He is currently the president of the GNU Radio Project and is actively involved in the effort to develop GNU Radio 4.0.

BRIEF CONTENTS

CONTENTS IN DETAIL

11
SDR HARDWARE UNDER THE HOOD

12
PERIPHERAL HARDWARE

13
TRANSMITTING

INTRODUCTION

There's a revolution going on in wireless communications technology, touching fields from hardware development to information security and from aerospace engineering to reverse engineering. Professional electronics designers, cybersecurity researchers, hardware hackers, and ham radio operators alike can all reap the benefits of the beautiful new combination of software and radio.

Since Marconi's first transmissions over 100 years ago, nearly all radios have been fixed-function devices. Any given radio could only tune to a relatively narrow range of frequencies and transmit or receive a specific kind of signal. At best, you'd have combination devices, like those that receive both AM and FM signals, but this was largely a matter of combining two (or more) fixed-function radios in one package. These old-school devices are *hardware-defined radios*. What defines the frequencies one of these radios can transmit or receive? Hardware. What defines whether it's an FM radio or a Wi-Fi router? Hardware. Nearly every aspect of these radios is defined by stubbornly inflexible hardware.

On the other hand, *software-defined radios (SDRs)* provide programmability throughout the radio's architecture. When you want to change the operation of an SDR, you don't have to rewire your hardware; you simply modify the programming on the SDR board and, quite often, on a connected host computer.

This book will show you how to build wonderfully flexible analog radios using an SDR and a program called GNU Radio, and it introduces just the right amount of basic theory to understand how those radios work.

A Week in SDR

It may not be immediately obvious how powerfully flexible an SDR can be. To illustrate, here's a story about a week in your life as an SDR user.

It's Monday, and you want something to talk to a number of Wi-Fi devices. You put together a project on your computer, connect to your SDR, and, sure enough, you have a Wi-Fi access point. And not just any access point, but one where you can control aspects of the channels used and the packets sent.

On Tuesday you need some geographical data from a job site you're visiting. With nothing more than some software changes, you reprogram your SDR to be a GPS receiver. As before, you're not just limited to basic GPS functionality but can keep track of how many satellites in the GPS constellation are observable and the signal strength of each. On the way back from the job site, you program your SDR to pick up FM broadcast stations so you can listen to a bit of music while you drive.

Sometime on Wednesday you realize that your Wi-Fi system is having some unknown difficulties. Suspecting security issues, you reconfigure your SDR to scan all 14 available Wi-Fi channels simultaneously and see what kind of traffic is out there. You save all the raw *radio frequency (RF)* data to a file for further processing at your leisure.

You shift gears on Thursday, turning your attention to some signals you saw broadcasting in the 2.4 GHz Wi-Fi band the day before. Not sure what they are, you start breaking down the signals using powerful but free and easy-to-use software. The mystery signals aren't currently transmitting, but that's not a problem because you have the raw data from yesterday. After looking at the data, you're able to determine that there's a ZigBee home automation network nearby, as well as a baby monitor and a poorly shielded microwave oven.

Now comes Friday, and you decide to drive around and check out the RF activity at various points in your city. Taking advantage of your Wi-Fi and GPS routines from earlier in the week, you're able to write a position-aware data logging application in just a couple dozen lines of Python code.

Finally, it's the weekend, and you decide to take a camping trip. Before you go, you reprogram your SDR to several different amateur radio modes in hopes of making contact with other ham operators while you're up in the mountains. Perhaps you might even ping the International Space Station!

Are you starting to see how powerful these devices are? Not only can you easily implement new radio designs, but you can also switch between them with nothing more than a few keystrokes or a function call in your code. More than simply transmitting and receiving, you can scan, find, and deconstruct other signals that may be out there.

In the interest of full disclosure, there's one caveat to all this: as programmable as SDRs have become, their creators still haven't figured out how to make affordable, programmable antennas. For now, at least, you may still have to swap those out when making big frequency changes.

SDR and Hardware Development

There's another key application for SDRs: prototyping new products containing an RF component. In the early stages of product development, it's critical to iterate quickly on new designs. Fail fast, and try as many new ideas as reasonably possible.

SDRs allow you to implement the radios in your design much faster than designing them from off-the-shelf components. You can also modify their functionality far more quickly. Due to cost constraints, you most likely won't take a design into high-volume production while it still contains an SDR, but you can optimize for cost later in the design cycle.

We've seen this before. For decades, digital logic was typically a hard-wired affair. You'd grab a bunch of chips containing logic gates, wire them together, and voilà: there's your system. If performance or cost demanded, you could even have a custom chip made (they're called *application-specific integrated circuits*, or *ASICs*). Then a new technology came and changed everything. Engineers started prototyping with different types of programmable logic, going by questionably helpful abbreviations such as PLAs, PALs, and PLDs. The most commonly used form today is the *field-programmable gate array (FPGA)*.

An interesting development occurred along the way. In certain cases, engineers started realizing that the cost of the programmable solution wasn't so much higher than the fixed-function solution after all. In fact, when factoring in the engineering hours required for designing the fixed-function implementation, sometimes the programmable solution was cheaper. And there was also the added benefit of being able to update the product's hardware functionality at any point in the production process and beyond—even when the product was in the hands of customers.

Although this little history lesson has focused on digital logic, similar programmable technologies now exist for analog circuitry too. One could even consider 3D printers to be in the same vein: *software-defined matter*.

All of these technologies reduce design cycle times and get products out more quickly. And while the earliest SDRs, FGPAs, and the like were primarily used for prototyping, their more mature descendants are increasingly finding their way into released products. This happens primarily in two circumstances. First, when time to market is critical, the additional cost of incorporating one of these technologies may be acceptable. Second, when you're not making a large quantity of your product, it doesn't make sense to spend the engineering hours to optimize its cost.

How long will it take for SDRs to become ubiquitous in the marketplace? It's hard to say for sure, but you might want to search for "RTL-SDR." In the past, this particular SDR design found its way into a number of consumer products (mostly digital TV tuners), while radio hobbyists hacked these products to turn them into extraordinarily low-cost devices for their own experimentation.

This Book's Approach

The goal of this book is not only that you'll learn about SDR but also that you'll have fun doing it. The word *fun* comes from *fundamentals*; learning how something works (for instance, a radio) makes that thing not only more useful to you but also more fun to use. Learn the fundamentals of a topic, then leverage them to have fun.

This book is designed for you to learn by doing. There won't be a ton of pages of dry exposition on electromagnetic theory and the mathematical underpinnings of signal processing. We don't believe many beginners to SDR are well served by a deep, formalized dive into these topics. To be sure, the book *will* cover these topics, but in a different way than you would encounter in an academic textbook: we'll start with the simplest possible concepts, pair them with actual SDR experiments, and then gradually build on what you've learned to go progressively deeper. The intent is to provide you with a functional understanding of the terms and concepts required to build practical SDR systems, and we'll always endeavor to not just tell you those terms and concepts but actually show them to you hands-on.

Think of SDR as an onion. Each chapter of this book will peel back a very thin layer of that onion (hopefully without any tears). Don't worry if you don't fully grasp a topic in an early chapter; the concepts will be fleshed out in greater detail as the book goes along and you work through more projects. After a number of chapters, we think you'll be pleasantly surprised at how deeply you've traveled into the heart of the onion.

There's one catch, however: learning by doing requires that you actually do the doing. Take the time (just a few minutes) to install the necessary software, then follow along and work through the book's examples so that you can actively start playing with radios, both simulated and real.

What do we mean by *simulated* radios? It turns out that the primary SDR software, GNU Radio, has the ability not only to control SDR hardware but also to simulate the operation of a real radio without hooking up any hardware at all. This simulation capability is the key to how we'll incrementally journey to the center of the SDR onion. Starting with simulations will allow you to grasp ideas more quickly, without the complications of integrating hardware and grappling with real-time radio data. Don't worry, though— we'll get to hardware in later chapters.

Who This Book Is For

This book is written both for those who have no SDR experience and those who have previously struggled to get off the ground with SDR. You may be a tinkerer, an amateur radio enthusiast, or a student. Alternatively, you may be an engineer who's forgotten most of the radio theory you learned in school and want a refresher. We just ask that you bring your curiosity to the subject.

The internet is useful for looking up information on just about any topic, but sometimes it can take a while to assemble data from a vast array of different websites into one cohesive mass of knowledge. We've had many people tell us they could search the internet to find software for running SDR hardware, but it wasn't clear how the various components worked. There was either "too much math" describing various SDR functions or too little practical information. If that kind of frustration sounds familiar, this book is for you. It won't eliminate the need for you to do your own research as your ambitions grow, but it will provide some of the fundamentals to better grasp the fascinating world of SDR.

This book's focus is on analog radios only. Building analog radios for receiving broadcast signals or working with amateur radio is what many of you reading this book will want to do, but the concepts covered are also crucial for those of you interested in digital applications, like reverse engineering or information security. Digital radios are built on the same foundational concepts as analog radios, so solidifying those concepts through simple analog applications will help you transition to digital communication.

What You'll Need

For the hardware-based activities in this book, we recommend using the HackRF One SDR. It's widely available, relatively affordable compared to many professional-grade SDRs, and well supported by GNU Radio and other open source software. If you possess a different SDR, such as a PlutoSDR or LimeSDR, don't worry: Chapter 9 outlines how to adapt the hardware projects in the book to work with other common devices. You'll also need an antenna that you can connect to your SDR. For the purposes of this book, we recommend the ANT500, but another antenna that can work with FM broadcast signals will also do. Many SDRs are sold bundled with a compatible antenna.

That said, you can start learning about SDR right away even if you don't have any hardware yet, since this book begins with simulated radios. Just head over to Chapter 1 and start reading while you wait for your new SDR to arrive. Because you can do so much with GNU Radio's simulation capabilities (and the input files we provide), you can actually learn quite a bit with nothing more than your computer.

What's in This Book

This book consists of three main parts. Here's a breakdown of what you'll learn in each section.

Part I: Building a Basic Receiver teaches you just enough theory and software to create your first SDR. You won't use any SDR hardware yet but rather will start out in a simulated environment. You also won't necessarily fully understand all the components of your radio and how they work at this stage, but this first part lays the groundwork for deeper dives in later chapters.

Chapter 1: What Is a Radio? Defines what a radio system is at the most basic level. You'll learn what radio signals are and see how radio systems use modulation and demodulation to transmit meaningful information with those signals.

Chapter 2: Computers and Signals Explores how radio signals are sampled and digitized so that your computer can store and process them. We'll discuss analog-to-digital and digital-to-analog conversion and highlight the importance of the sample rate in these processes.

Chapter 3: Getting Started with GNU Radio Introduces GNU Radio and its graphical user interface, GNU Radio Companion, the software you'll use throughout this book. You'll install and test the software and learn about its block-based interface for creating visual programs called *flowgraphs*.

Chapter 4: Creating an AM Receiver Walks you through the process of creating a basic AM radio receiver using GNU Radio. You'll test out your receiver in a simulation by feeding it a file of previously captured real-world radio data.

Part II: Inside the Receiver gradually unpacks the concepts and components behind the radio built in Chapter 4, giving you a deeper understanding of how radios work and showing you how to build even better radios in the process.

Chapter 5: Signal Processing Fundamentals Takes a deeper dive into three foundational concepts in signal processing: frequency, gain, and filters. Through hands-on experiments, you'll learn about the audible spectrum, view the frequency components of signals, apply gain to strengthen and attenuate signals, and use a variety of filters to isolate different parts of a signal.

Chapter 6: How an AM Receiver Works Circles back to the AM receiver from Chapter 4. Armed with your new knowledge of signal processing, you'll take a closer look at each part of the receiver to understand how it works and why you configured it the way you did. You'll also learn how the receiver is able to tune to different signals.

Chapter 7: Building an FM Radio Shows you how to adapt your AM receiver to work on FM signals (still simulated). In the process, you'll learn several tips for building cleaner and more powerful radios.

Part III: Working with SDR Hardware leaves the world of simulated radio mostly behind and guides you through integrating SDR hardware with your GNU Radio flowgraphs. You'll find out how your SDR hardware works and gain a still-deeper understanding of signal processing, including how signals are sent and received.

Chapter 8: The Physics of Radio Signals   Fills in more of the gaps on the properties and propagation of radio signals. You'll learn about electromagnetic waves, understand the bandwidth of your signals, and explore the ramifications of noise on your radio data.

Chapter 9: GNU Radio Flowgraphs with SDR Hardware   Walks through the steps to modify your software FM radio from Chapter 7 to work with real SDR hardware. You'll finally test out your receiver on radio data you capture yourself, in real time.

Chapter 10: Modulation   Details the three basic kinds of modulation: amplitude, frequency, and phase. You'll work through experiments demonstrating each type as you learn how to avoid overmodulation, adjust your modulator's sensitivity, and more.

Chapter 11: SDR Hardware Under the Hood   Pulls back the curtain on how SDR hardware is able to grab signals over the air and send them to your computer (and vice versa). You'll learn about IQ sampling, revisit concepts like analog-to-digital conversion and gain, and get to know some important specs to look for in an SDR.

Chapter 12: Peripheral Hardware   Discusses some of the practical concerns around using SDR hardware, including what kind of antenna you should use, what kinds of connectors and cables you need, and how to mitigate noise.

Chapter 13: Transmitting   Outlines how to build an SDR transmitter using FM modulation and highlights some of the practical and legal issues behind transmitting. You'll test your transmitter in a simulation by connecting it to an SDR receiver.

Online Resources

To make your SDR learning experience as smooth as possible, all the necessary project and input data files can be downloaded from this book's web page at *https://nostarch.com/practical-sdr*. Extract the contents of the download into a convenient location on your hard drive because you'll be using the files frequently.

The contents of the compressed file are broken down by chapter, so you can easily find the files you need as you work through the book. You'll get the most out of the book if you build each project from scratch as shown in the book's text, but the finished files are available as a reference in case you run into trouble.

Each chapter typically contains one or more input data files in addition to the completed projects. Thanks to these input data files, it's possible to work through most of the material in this book without needing access to any SDR hardware.

PART I

BUILDING A BASIC RECEIVER

1

WHAT IS A RADIO?

In this chapter, we'll consider a very simple model of a radio system. We'll then expand on that model by unpacking some of its underlying concepts. By the end of the chapter, you'll have a general idea of what a radio is, what radio signals are, and how modulation and demodulation allow you to transmit information using radio signals.

A Simple Radio Model

At its simplest, a *radio system* is a pair of magic boxes that physically communicate through the air, without wires.

Consider a real-world example. You've probably played with a car radio at some point. It picks up transmissions sent out into the air by a big radio tower somewhere in the vicinity and turns those transmissions into sound. In other words, this radio system includes a transmitter (the big tower), a receiver (the car radio), and a transmission medium (air), as shown in Figure 1-1.

Figure 1-1: A very simple radio model

As you can see, the transmitter is producing something and sending it through the air to the receiver. But what's that "something" moving through the air? It's a signal.

Signals

A *signal* is what happens when we change some kind of physical property to convey information. Let's consider some examples.

Imagine for a moment that you and a friend are working on the opposite sides of a large, flat field. Also imagine that it's nighttime and that every so often your friend needs your assistance. In this rather contrived scenario, your friend needs a way to communicate that you should cross the field to help out. Imagine they do this by turning on a bright flashlight and pointing it across the field toward you. When you see the light in the darkness, you know it's time to come help.

Don't ask why you don't just use cell phones. That messes up the example.

In this simple thought experiment, your friend's flashlight is the transmitter, and your eyes are the receiver. Turning on the light generates a change in a physical property (the light level) that your eyes (the receiver) can interpret as information ("Come on over!"). Thus, when your friend turns on the flashlight, they're sending you a signal.

Another example of a signal is a message on an old-fashioned telegraph system. The transmitter and receiver are connected by a wire. On one end of the wire, a transmitter alternates between applying electricity to the wire and then removing the voltage. On the other end of the wire, a receiver creates a beeping tone when it senses an electrical current and is quiet when the current goes away. Using Morse code, a system in which letters and numbers are represented by different patterns of beeps and silences, the telegraph operators can send text messages to each other. In this case, the signals are based on the changing physical property of electric current running through a wire.

Since this is a software-defined *radio* book, we'll be most interested in radio signals. You might recall from science class that radio signals are electromagnetic in nature. This means that radio transmitters represent information by changing the properties of electromagnetic waves. Don't worry, though: you don't need to get into much math or physics to start working with SDR.

What information can you communicate using radio signals? It could be several things:

- Sound, like the music or speech in an AM or FM radio broadcast
- Video, such as over-the-air television broadcasts

- Control information, like the unlock command sent from your wireless key fob to your car
- Data, such as the web traffic going between your Wi-Fi router and your laptop
- And so much more!

Since these information types are themselves quantities varying over time, each of them is a signal in its own right. Radio signals must then have a clear relationship to the information contained within them. To understand how radio signals can carry all these different types of information, you'll need to understand a little about modulation.

Modulation

Modulation is what makes it possible to transmit meaningful information via radio signals. Essentially, modulation is a way of combining two signals together by using some property of one signal to change some property of the other. The first of these two signals is simply the information we're trying to communicate. It modulates, or changes, a second signal called the *carrier*. The output of this modulation process is a signal that has characteristics of both input signals, ready for transmission via a radio system.

It's also possible to take a signal that has gone through the modulation process and separate it back into its two original parts. This reverse process is called *demodulation*. It's what happens at the receiving end of a radio system to turn a radio transmission back into meaningful information.

A Slightly More Complicated Radio Model

We now have enough information to return to our basic model of a radio system and fill in some of the details. This time, we'll think more specifically about a car's AM radio dial. In fact, AM radio is the simplest kind of radio around, and it's a topic we'll return to throughout the book.

Remember that our basic radio model began with a big transmission tower. Imagine that at the site of the transmitter, there's some audio that the radio operator wants to broadcast. It could be music or someone's voice. The operator uses this audio signal to modulate a carrier signal assigned to the radio station and then sends the resulting modulated signal out through the air (there are some significant simplifications here). For the final step in this process, a user turns the AM dial on their car radio to tune to the desired station (hint: this tuning process will have something to do with that carrier signal) and demodulates it to recover the original audio. This process is shown in Figure 1-2.

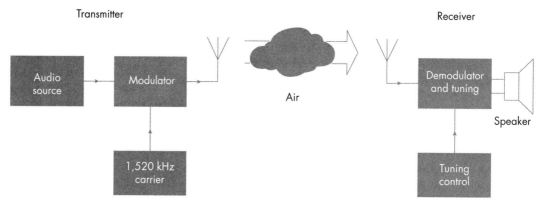

Figure 1-2: A less simple radio model

Notice that the carrier signal in the diagram is labeled 1,520 kHz. Short for *kilohertz*, kHz is a unit of frequency, and this 1,520 kHz corresponds to 1520 on your AM radio's dial. We'll discuss frequency, one of the fundamental concepts in signal theory, in more detail later. For now, just consider that the music coming out of your car radio hints at an important concept: radio engineers have found a way to take sound and send it out to the world on different radio frequencies.

More specifically, at the various AM channels on your dial, you have sound somehow being transmitted, with different folks transmitting at different frequencies simultaneously. In Seattle, for example, there's audio flying around at 570 kHz, 1,150 kHz, 1,300 kHz, and many other AM frequencies. All of the radio towers in the area are sending out their signals all the time, but somehow it's possible for you to listen to just one of them by turning a knob. How is that?

Part of the answer is that radio receivers are capable of *tuning*, or focusing in on just one signal at a time. Tuning is one of the most important skills you'll learn in this book, so rest assured that we'll cover how it works in depth before we're done. In the meantime, let's take a closer look at the actual signals involved in AM radio broadcasts. This will get us closer to understanding how a radio station can send sounds out into the world.

AM Radio Signals

As mentioned before, you can generally think of a signal as the variations in some physical property over time to reflect information. Maybe it's variations in air pressure that make up the sounds reaching your ears. Maybe it's the voltage on a wire that goes to your radio's speaker. Maybe it's the electromagnetic intensity received by a radio antenna. Despite the wildly different underlying physics of these signals, they all look similar on paper: they'll each have some kind of vertical axis representing the signal's value and a horizontal axis representing time. For example, Figure 1-3 shows what the signal for a simple audio tone looks like when plotted on x- and y-axes.

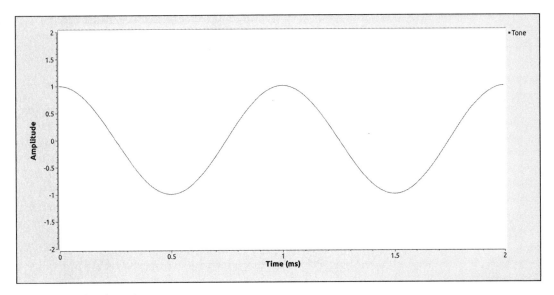

Figure 1-3: A simple audio tone

As time progresses, the signal in Figure 1-3 just keeps oscillating back and forth for as long as the tone persists. You might recognize its shape from a trigonometry class as a *sine wave* or *sinusoid*. Let's assume that the AM radio station operator wants to transmit this very simple sound.

Now let's look at the AM station's carrier signal, shown in Figure 1-4. It's actually similar to the simple audio tone; it just moves back and forth a lot faster.

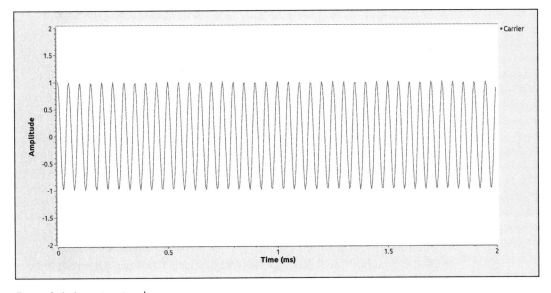

Figure 1-4: A carrier signal

We now have an audio signal we want to send and a carrier frequency that the radio station operates at. How do we send the audio signal using the carrier signal?

Amplitude Modulation

This is probably a good time to mention that the *AM* in *AM radio* stands for "amplitude modulation." *Amplitude* refers to the strength of a signal (the property shown on the y-axis of the signal diagrams you've been looking at), so *amplitude modulation* means that we modulate, or change, the strength of the carrier based on the strength of the audio signal. Specifically, we proportionally reduce the strength of the carrier when the audio signal is low and increase the carrier's strength when the audio signal is high. Figure 1-5 shows a modulated signal based on the audio tone and carrier from the previous two figures.

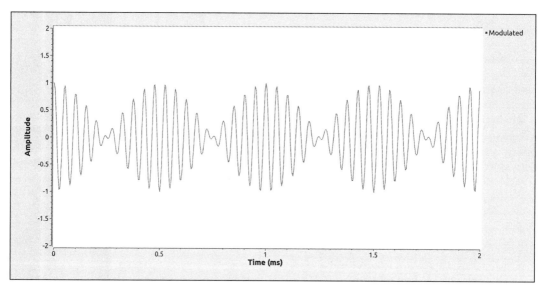

Figure 1-5: The carrier amplitude modulated by a simple tone

Notice first how the peaks and troughs of the modulated signal are the same distance apart as the peaks and troughs of the carrier signal shown in Figure 1-4. In this sense, the modulated signal is still closely related to the carrier. But notice also how the heights of the signal's peaks have changed to take on the shape of the audio tone shown in Figure 1-3. You can actually see the original audio signal imprinted on the shape of the carrier. That's amplitude modulation.

You might be wondering what this modulation process looks like for a real-world audio signal, like a person talking or a band playing music. After all, radio stations rarely just broadcast a simple tone. These more interesting kinds of audio signals are much more complicated, as you can see in Figure 1-6.

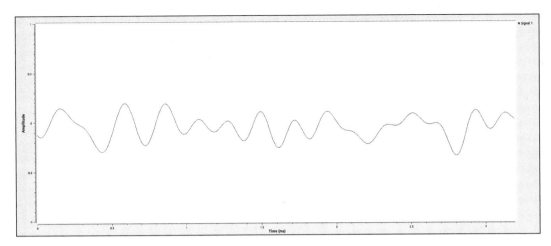

Figure 1-6: A real-world audio signal

Even though this audio signal is more complicated than a simple musical tone, modulating it onto a carrier produces the same result: you can still see the outline of the original audio signal on the modulated carrier, as shown in Figure 1-7.

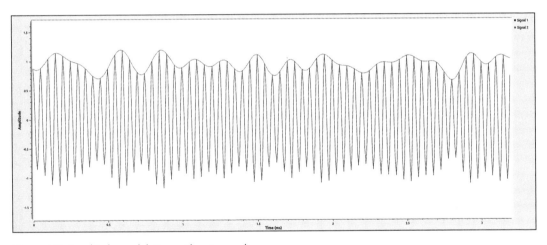

Figure 1-7: Amplitude modulation with voice audio

Demodulation simply reverses this process, allowing us to take a modulated signal and extract the signal originally used to modulate the carrier.

Conclusion

You now have a very basic idea of what signals are, and you've seen how modulation can use an audio signal to change the shape of a carrier signal. You also know that demodulation can "undo" this. Later you'll work on actually building modulators and demodulators using GNU Radio, so you can see how modulation and demodulation fit into the structure of SDR transmitters and receivers.

There are still a lot of unanswered questions: Why do we even need a carrier? How do we choose the carrier frequency? What other kinds of modulation are there? We'll get to all those questions and more, but first you need to understand how to work with signals using a computer.

2

COMPUTERS AND SIGNALS

The signals we examined in the last chapter all looked like squiggly lines moving around over time. This chapter answers the question, "How do I get those squiggly lines into my computer so my software can work with them?" We'll take a little excursion into the realm of *digital signal processing*, a whole field of study dedicated to using computers to capture, manipulate, and reproduce real-world signals. In particular, we'll look at *sampling*, the process of repeatedly measuring a signal.

We'll only scratch the surface of digital signal processing in this chapter, but that's okay: you can start learning about SDR hands-on with just a few concepts from this field under your belt.

Digital Sampling

Sampling is really just measuring. Computers need numbers to work with, and the sampling process consists of a series of measurements over time to convert those squiggly, real-world signals into numbers. If you make those measurements quickly enough and accurately enough, you'll get a series of numbers that reasonably represent the original signal. We call each of those measurements a *sample.*

Consider a signal in the form of a voltage on a wire. What we need to do is translate that voltage into a digital value that a computer can understand. In other words, measure it! However, the voltage might change over time: it could be 1.3 volts (V) one moment, −0.042 V another moment, and 110 V the next. To accurately represent this changing signal, we need to keep measuring it (that is, keep taking samples), generating new digital values from those measurements for as long as we want to look at the signal.

Analog-to-Digital Conversion

The hardware that repeatedly samples a signal and translates those samples into computer-readable values is called an *analog-to-digital converter,* or *ADC* for short, because the real-world signals are considered *analog* and the numerical measurements of them are considered *digital.*

Let's look at an example to see how this works. Possibly the simplest signal you'll ever encounter is a *square wave.* It alternates periodically between two different values. In the case of the square wave shown in Figure 2-1, it switches from 0 V to 3 V and back to 0 V, and it does so every second.

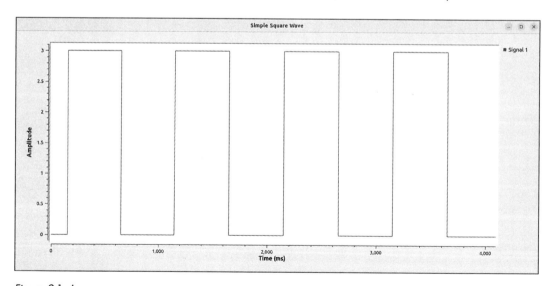

Figure 2-1: A square wave

When we feed this signal to an ADC, the ADC will take a series of measurements, producing a steady stream of samples. Each of these samples will be valued at 0 or 3, representing the voltage level of the signal at the moment each sample is taken. Figure 2-2 shows a plot of the output samples, with each sample displayed as an individual dot.

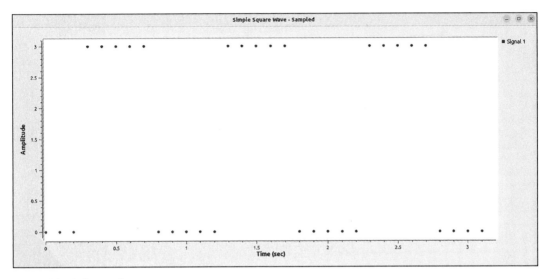

Figure 2-2: A digitized square wave

As you can see, we have a series of dots, or samples, unfolding horizontally across the diagram. Some of the dots are at 3, and some are at 0. Connect those dots together, and you get a pretty good representation of the original signal. Meanwhile, a computer would receive these samples as a data stream or array with the following contents:

```
[0, 0, 0, 3, 3, 3, 3, 3, 0, 0, 0, 0, 0, 3, 3, 3, 3, 3, 0, 0, 0, 0, 0, ...]
```

Now take a look at a sinusoid and its sampled version in Figure 2-3. You may recall the discussion of sinusoids in Chapter 1. They're an important waveform type with unique properties that are useful for SDR work. Get used to seeing them.

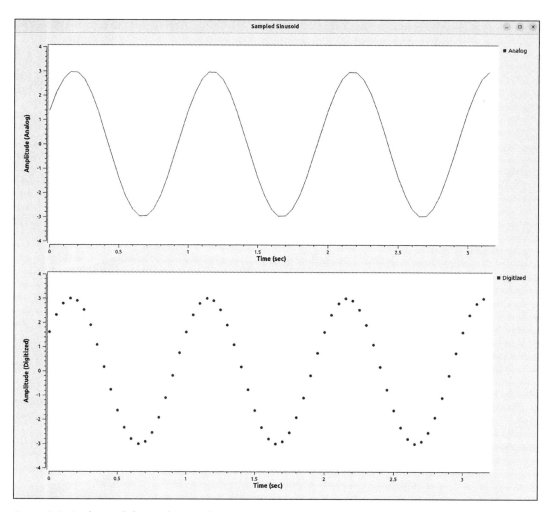

Figure 2-3: Analog and digitized sinusoids

The smooth oscillations of the sinusoid are a good illustration of the continuous nature of analog signals, which an ADC breaks up into a collection of discrete samples. Even so, look at the dots in the lower half of the figure, and you can still kind of make out the shape of the original sinusoid.

Just so you don't think ADCs are limited to artificial signals, Figure 2-4 shows a more arbitrary waveform, similar to the audio snippet you saw in Chapter 1.

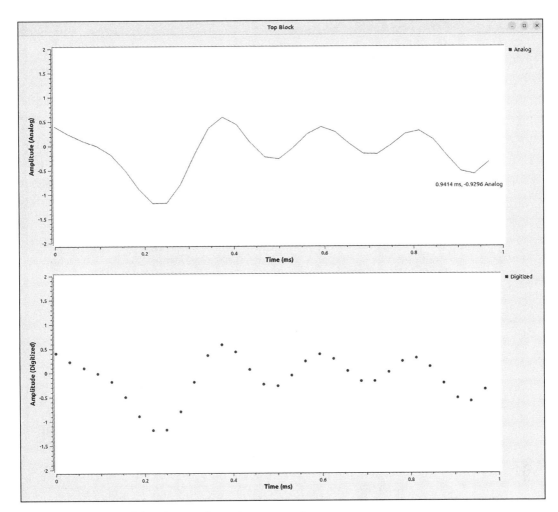

Figure 2-4: Analog and digital views of an arbitrary waveform

In this case, the number stream the computer receives would be roughly as follows:

```
[+0.4, +0.2, +0.1, 0, -0.2, -0.5, -0.9, -1.2, -1.2, ...]
```

While you can still more or less see the original arbitrary waveform in the sampled version, some of the nuances of its shape have been lost. We'll discuss why in a moment.

Digital-to-Analog Conversion

We can also flip the ADC process around and convert digital (computer) values to analog (real-world) values. The device that does this is unsurprisingly called a *digital-to-analog converter*, or *DAC*. A DAC is what makes it possible, for example, to plug in your headphones and listen to a song stored on your computer's hard drive. The DAC takes the digital information from the audio file and converts it into a real-world signal that can be played through your headphones.

In summary, an ADC measures signals in the real world and translates them into a form your computer can understand. The DAC does the opposite, allowing you to take the values your computer comes up with and send them out as a real-world signal. As you probably expect, a lot of details are missing in this simple description. We haven't discussed precision, noise, anti-aliasing filters, binary storage, and so on. We'll get to some of these concepts in later chapters. The one thing we do want to cover a bit more at this stage, though, is the sample rate.

Sample Rate

The *sample rate* (sometimes called *sampling rate*) is the number of times per second you take samples (that is, make measurements) of a signal. Conversely, the amount of time that passes between each individual sample is called the *sample period* (or *sampling period*). Take a look back at the sampled signal diagrams we just discussed. The horizontal distance between one dot, or sample, and the next represents the sample period. Count up the number of dots per second, and you get the sample rate.

Think of the sample rate as how fast you tell the ADC to make measurements of the signal. The higher the sample rate (and the shorter the sample period), the better your computer representation of the signal will be. On the other hand, the more samples you take per second, the more data you'll end up with, meaning you'll need more storage space and processing power to handle it. Also, your hardware will be limited with respect to how fast you can sample.

The most important question, then, is "How fast should I sample my signal?" To answer, first look at Figure 2-5, where a sinusoidal waveform has been sampled at a relatively high sample rate.

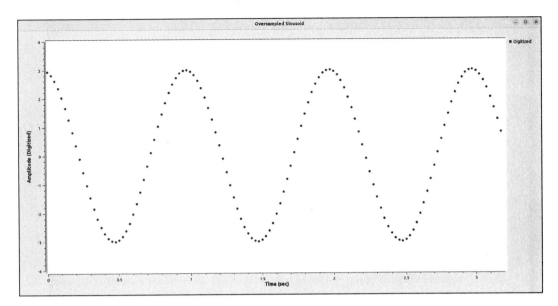

Figure 2-5: A digitized sinusoid with a high sample rate

See how the original continuous signal is visible even to the naked eye? When your computer receives this data stream, it will have a faithful reproduction of the original signal. In fact, the sampling in Figure 2-5 is actually many times faster than necessary for this particular signal, but we've exaggerated it for clarity of visualization.

On the other hand, Figure 2-6 shows what happens when we sample much more slowly. Compare the continuous waveform (top) with the sampled one (bottom).

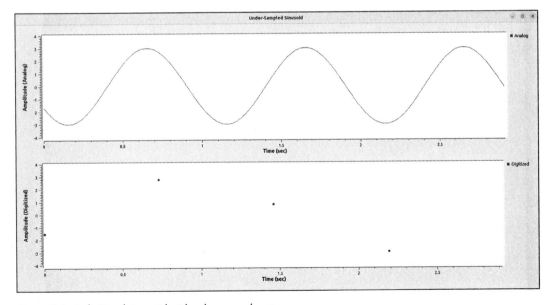

Figure 2-6: A digitized sinusoid with a low sample rate

You can't really see the original signal anymore, can you? It's because we're sampling so slowly that we're missing things. By *slowly*, we don't mean that the sample rate is slow in an absolute sense, but that it's slow relative to the signal we're trying to sample. The sample period is so long that the quickly changing signal moves around too much between successive samples. In between each pair of samples, we essentially lose track of the signal.

Here's one way to think of it: Imagine you're standing in a room with a very active cat. It likes to pace around in a big circle, but it somewhat randomly changes direction. For the sake of illustration, let's say the cat can stroll through a complete circuit of the room in 10 seconds at its very fastest. But sometime within those 10 seconds, the cat might slow down or turn around and start strolling the other way; in fact, it might change speed or direction several times.

Now imagine that someone turns off the lights. Then that same someone starts flipping the light switch up and down. Every second, they flip it on for an instant and then back to darkness. The result of all this is that every second you get a brief glimpse of where the cat is located. In other words, you have a sample rate of one sample per second. Since the light is flickering fairly quickly relative to the slow cat, you have a pretty good idea of the cat's motion. Even when it changes direction or speed, you can be sure you know where the cat is and where it has just been. It just doesn't move fast enough to escape being observed in the flickering light.

Next imagine that the time between light flickers increases significantly. Now the light switch only toggles to give you a view of the room every 60 seconds. Under these conditions, do you think you'd have any idea how the cat is moving? In between those rare flashes of light, the cat could be doing anything: switching direction, speeding up, slowing down. When you "sample" the room too slowly relative to the speed of the cat, you have very little idea of what the cat is doing.

A similar phenomenon is happening with our *undersampled* sinusoid in Figure 2-6. We're sampling it too slowly relative to the speed of the signal, and we're missing changes in the signal as a result.

Here's the most important takeaway from this discussion: sample fast enough, or you'll get bad data. How fast is fast enough? Answer: you need to sample significantly faster than the signal you're trying to measure.

Assuming the signal isn't a simple sinusoid, how do we know how fast it is? Another great question. Don't worry, we'll address this in the coming chapters.

SDRs from 50,000 Feet

Now that you're armed with a basic understanding of analog-to-digital and digital-to-analog conversion, you can begin to understand how an SDR works. Figure 2-7 shows a simple model illustrating how an SDR and a computer can receive radio signals.

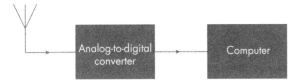

Figure 2-7: A simple SDR receive model

Your receiving antenna picks up some radio signals, an ADC translates them into a stream of numbers your computer can understand, and the computer then processes the number stream to make sense of the signals. On the transmit side, your computer generates a digital version of a signal, a DAC translates the signal to analog, and then the transmitter portion of the SDR sends that signal to the world through the transmit antenna. Figure 2-8 gives you a view of that.

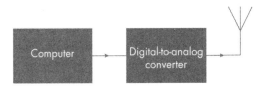

Figure 2-8: A simple SDR transmit model

These models leave out *a lot* of details, but they're a good starting point. The key is that we have a computer at the root of both block diagrams, one that you can reprogram to do almost anything you can imagine. You can:

- Extract audio from radio signals and play it back through your speakers
- Capture raw radio data to a file so that you can analyze it later
- Change the SDR's programming so that you can transmit or receive completely different types of signals whenever you need to
- Find mysterious signals and reverse engineer them

This list only scratches the surface of what your computer can do in tandem with an SDR and digital sampling.

Conclusion

In this chapter, you learned how an ADC can take a real-world signal and translate it into information a computer can use. It does this by taking lots of individual measurements, or samples, of the signal. You also learned that a DAC performs a similar process in reverse. You discovered the importance of the sample rate: by sampling fast enough, you ensure that you capture the faster-moving parts of a signal. Finally, you saw how ADCs and DACs are essential to the world of SDR: they allow your computer to take in and send out radio signals.

But enough theory. It's time to start learning by doing! In the next chapter, we'll start digging into GNU Radio, the powerful software we'll use for capturing and processing radio signals using SDRs. GNU Radio will lie at the heart of the computer block in this chapter's block diagrams.

3

GETTING STARTED WITH GNU RADIO

It's finally time to start playing with GNU Radio, a free and open source collection of software capabilities that support radio design graphically as well as through traditional text-based coding. In this chapter, you'll install GNU Radio and test it out by creating and running a few simple *flowgraphs*. These are programs built using GNU Radio's graphical user interface, GNU Radio Companion. You'll learn how to get data into and out of your flowgraphs and how to manipulate that data, all using a simple, block-based interface. You'll see how GNU Radio lets you create new radios almost entirely in software.

Installing GNU Radio

Once upon a time, and by that we mean way back in the 2010s, installing quality SDR software required quite a bit of time and effort. Fortunately, those days are behind us, and due to the inexorable progress of open source software and the incredible efforts of its contributors, you'll be up and running with GNU Radio in no time.

First, a word about your computer's operating system. For many years, the best advice for running GNU Radio was to use Linux, the operating system (OS) on which GNU Radio was developed. Installation on Linux was easier and more dependable than on systems running another OS. Fortunately, there are now much more reliable ways to install GNU Radio on Windows and macOS computers. The installation process differs depending on your OS, so we'll discuss each option separately.

Linux

The specific Linux distribution we recommend for installing GNU Radio is Ubuntu, and for the purposes of this book, we'll be using Ubuntu 24.04 LTS. GNU Radio will also work on numerous other versions and distributions of Linux, but we won't cover all of the possibilities here. You can find installation help for other distributions on the GNU Radio website (*https://www.gnuradio.org*).

To install GNU Radio in Ubuntu, simply open a terminal window and enter the following commands:

```
$ sudo apt update
$ sudo apt -y upgrade
$ sudo apt -y install gnuradio
```

You'll need to enter your password for the first of these commands if you haven't done so recently.

And that's it! You're now ready to start using GNU Radio.

Windows and macOS

The best way to install GNU Radio on Windows or macOS is to use radioconda. Rather than install the software natively, or directly, on your computer, radioconda creates a virtual environment containing everything you need to run GNU Radio. It's derived from Conda, a more general system for managing Python virtual environments.

To get started, download the latest radioconda installer from the project's GitHub page and execute it on your computer. Once installed, you simply need to switch from your computer's native environment to the virtual environment by activating radioconda. You can then run GNU Radio in this new environment and, if necessary, deactivate it when done. Don't worry if you haven't worked with virtual environments before; activating and deactivating are all you'll need to do.

The specific steps on how to install and use radioconda can change over time, so see the latest instructions on the GNU Radio website or visit *https://www.factorialabs.com/install*.

A Virtual Machine

Another option for running GNU Radio on non-Linux computers is to use a virtual machine (VM). With commercial software from companies such as VMWare or Parallels, or with open source software like VirtualBox, you can run an instance of Ubuntu virtually on your Windows or macOS machine. To do this, you first must create the VM per your virtualization software's instructions. Then, once you've logged into the Ubuntu VM, follow the instructions outlined in "Linux" on the previous page to install GNU Radio. This approach can be a bit trickier and can have a negative performance impact, so unless you have a strong reason for doing otherwise, we recommend starting with the radioconda approach.

GNU Radio Companion

Now let's take a moment to explore some of the key features of GNU Radio Companion, the graphic user interface (GUI) for GNU Radio. Rather than writing programs in languages such as C++ or Python, with a host of text files containing such things as if statements and while loops, GNU Radio Companion lets you build programs graphically by linking together different *blocks*. Because of the way data flows from block to block, these programs are called *flowgraphs*.

NOTE *The rest of this chapter is written from the vantage point of a Linux user. You'll need to enter the appropriate command for Windows or macOS (as shown on the install web page) if you're using either. Additionally, there will be minor differences in the appearance of the user interface on non-Linux systems.*

Start up GNU Radio Companion from a terminal window by entering the `gnuradio-companion` command.

A split second later, you'll see a window like Figure 3-1. We've annotated the figure to highlight the major parts of the GNU Radio Companion interface.

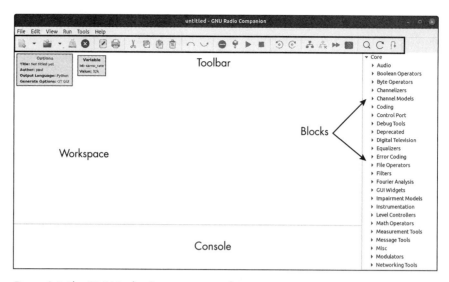

Figure 3-1: The GNU Radio Companion interface

The menu on the right side of the window lists all the blocks available for use in your flowgraphs. To actually make a flowgraph, you'll drag blocks from the list into the workspace, the main part of the window. Then you'll connect the blocks in whatever way you choose. Within the workspace, you'll also configure each of the blocks depending on the particular needs of your flowgraph. You'll use the icons in the toolbar across the top of the window to open, close, save, and run your flowgraphs. The console at the bottom of the window displays any warnings or errors.

Figure 3-2 shows an example of a flowgraph with several connected blocks. Throughout this book, you'll be working your way toward creating this kind of flowgraph.

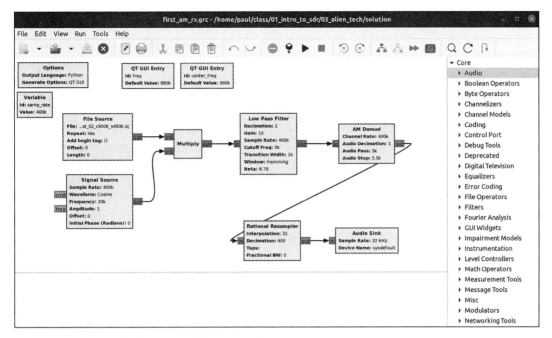

Figure 3-2: An example flowgraph with blocks and connections

Of the many different types of blocks available in GNU Radio Companion, perhaps the most important are sources and sinks. These are the blocks that allow you to get data into and out of your flowgraphs.

Sources

Source blocks insert streams of data into your flowgraph. Remember the analog-to-digital examples in the last chapter, where long sequences of numbers streamed out of the analog-to-digital convertor? You can inject streams of radio data just like that into your flowgraph, from your SDR hardware, using an SDR source block. This is perhaps the most important type of source, but you won't be working with those until later in the book. For now, you'll focus on several source blocks that don't require any hardware.

A `File Source` block, for example, gets its data from a selected file on your computer. For the purposes of this book, we've generated a data file that simulates the behavior of real SDR hardware. Throughout the first part of this book, you'll use `File Source` blocks to stream this data into your flowgraphs for processing. This way, you can learn the basics of SDR much more easily, without worrying yet about the nuances of the hardware.

Another key block you'll use is the `Signal Source`, which generates pure sinusoids as well as other synthetic signals and injects them into your flowgraph. The `Constant Source` is even simpler, creating a stream of numbers, each one identical to the next. You'll soon see why these kinds of source blocks are useful.

Sinks

The counterparts to the source blocks are *sink blocks*, the outputs of your flowgraphs. Data flows into your flowgraph via sources and exits your flowgraph via sinks. Sink blocks that send data to SDR hardware for radio transmission are a crucial type, but like source blocks, not all sink blocks require special hardware. For example, you can send data to an `Audio Sink`, which drives your sound card, to hear what it sounds like. Or you can capture the data your flowgraph produces and save it for later using a `File Sink`.

Although many of the sink blocks have source block counterparts, one special type of sink block does not. These are the *instrumentation blocks* that allow you to visualize your flowgraph's data in real time using a number of different techniques. In a moment, you'll create your first flowgraph, and it will include one of these instrumentation blocks as the output.

Hello, SDR!

As a "Hello, world!"–type introduction to GNU Radio, your first flowgraph will generate and display a constant signal. As you create this simple project, you'll learn a number of crucial concepts about how to build and run flowgraphs.

You should already have a blank flowgraph open in GNU Radio Companion, as this is the default behavior when you start it the first time. If not, create a new one by clicking **File ▶ New ▶ QT GUI** or using the CTRL-N keyboard shortcut. You're now ready to start creating your first SDR masterpiece!

Adding Blocks to a Flowgraph

It's possible to hunt through the complete list of 100+ blocks until you find the one you want, but it's much easier to find blocks using the search function, at least when you know part of the name of the block you're looking for. To access the search function, click the magnifying glass icon near the right end of the toolbar. It'll bring up a search box at the top of the list of blocks, as shown in Figure 3-3.

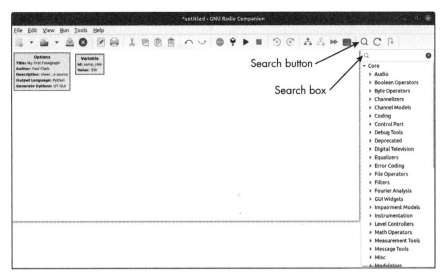

Figure 3-3: The search button and search box

Our goal for this flowgraph is to build a stream of unchanging numbers and then display them graphically. The Constant Source produces a stream of numbers with a constant value, so you'll start by adding one of these to your project. Click the search button (or use CTRL-F) and type **constant** into the search box. With each character you type, the block list is filtered accordingly. After you type a few characters, you'll see Constant Source at the top of the list. Click the block name and drag it into your workspace, as shown in Figure 3-4.

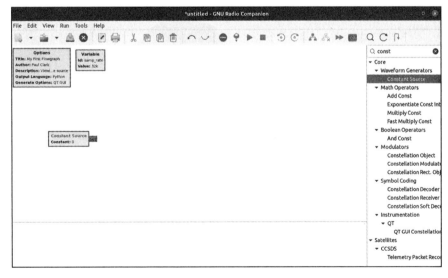

Figure 3-4: Adding the first block to the flowgraph

Although it's not necessary, it's a good idea to drag the Constant Source block to the left side of the workspace. By convention, flowgraphs are read from left to right, with source blocks appearing on the far left and sink blocks appearing on the far right.

Next, add a Throttle block to the right of Constant Source. For now, we won't get into the details of what this block does, but just know that it prevents your computer from running much faster, and consequently hotter, than necessary. Use your newly acquired search button skills to replace the search box text with **throttle**, and double-click the Throttle block to place it in the workspace.

The last block to add to your flowgraph is the QT GUI Time Sink. This is one of those instrumentation blocks that will allow you to visualize the digital data stream in your flowgraph. Modify the search box text to **qt gui time s** (capitalization doesn't matter) and select the block. Your workspace should now look like Figure 3-5 (note that from now on, we'll usually show only the workspace portion of GNU Radio Companion, not the entire interface in the figures).

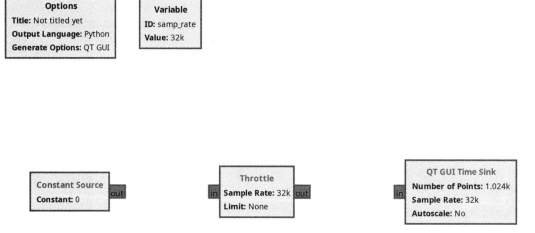

Figure 3-5: The flowgraph with three blocks

Don't worry if the blocks aren't in exactly the same positions as they are in the figure. Positioning won't affect how the flowgraph functions. If you *do* want to move your blocks around, however, you can do so by clicking and dragging them. While it won't affect functionality, it's a good habit to organize your flowgraph neatly so it's as easy as possible to see what's going on. Having the blocks haphazardly strewn about the workspace is a lot like writing source code with randomly indented lines: it's not great.

Connecting Blocks

The next step is to connect the blocks together so that data can flow from the source to the sink. First, connect Constant Source and Throttle. Click the port on the right side of the Constant Source block labeled Out, then click the port on the left side of the Throttle block labeled In. Did you see the connection appear? This is how you'll make all the connections in your flowgraphs: by clicking from the output port of one block to the input port of another.

Notice that the connection isn't just represented with a line. There's also an arrow to show the direction of data flow. In this case, the data is flowing out of the source block.

Finish your flowgraph by connecting the right port of the Throttle block to the sole tab of the QT GUI Time Sink. When you're done, you should have something like the flowgraph in Figure 3-6.

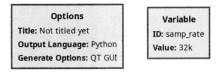

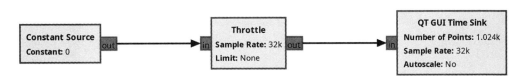

Figure 3-6: The completed flowgraph

You've just built your first flowgraph. Congratulations!

Saving and Running the Flowgraph

Just like you would any document you edit on a computer, you need to save the file containing your flowgraph. Click **File ▸ Save** or press CTRL-S and then save the file as *source_sink.grc* using the dialog that comes up. Finally, it's time to run the flowgraph. Click the **Execute** icon on the toolbar, as shown in Figure 3-7.

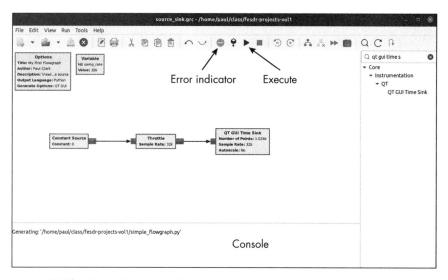

Figure 3-7: The Error indicator and Execute buttons

If there are errors in your flowgraph, the Execute button will be grayed out. Also, some part (or parts) of the flowgraph will be colored red (though they're shown in black and white in this book), which will help you figure out which block (or blocks) is the culprit and how to fix it. After a moment of computation, GNU Radio Companion will bring up a display window showing you a constant waveform with a value of 0, as shown in Figure 3-8.

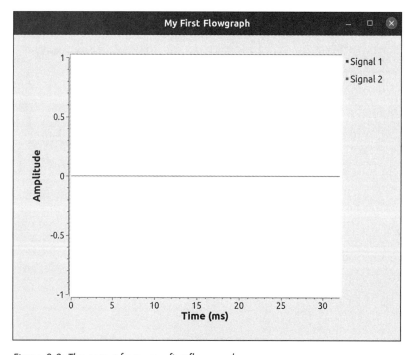

Figure 3-8: The output from your first flowgraph

Take a moment to gaze upon the beautiful waveform. If the display window is smaller than you prefer, feel free to click and drag a corner of the window to resize it. When you're ready, close the window by clicking the X in the upper-right corner. Closing the display window also terminates the execution of the flowgraph.

Changing Block Properties

A line at 0 isn't too exciting, is it? To make your flowgraph marginally more interesting, you can modify the Constant Source. To change a block's behavior, double-click the block and update its properties, so go ahead and double-click **Constant Source** and then change the value of the Constant property to 3.14. If it looks like Figure 3-9, click **OK**.

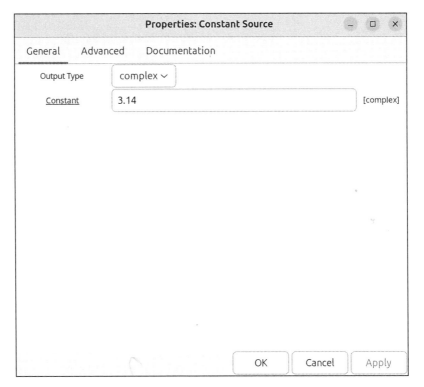

Figure 3-9: The Constant Source properties window

The flowgraph will now look the same as before, but with one exception. Notice how the rendering of the Constant Source block in the workspace has changed to reflect the new value, as shown in Figure 3-10. GNU Radio Companion blocks typically re-render in the workspace to show what properties are currently set within the block. This makes it much easier to see what your flowgraph is doing without having to click into each block to see its properties.

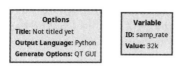

Figure 3-10: The modified flowgraph with a new Constant Source value

Execute the flowgraph again, and you'll see the window shown in Figure 3-11.

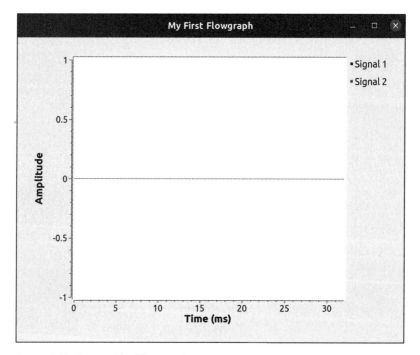

Figure 3-11: The modified flowgraph output

You modified the Constant Block, but the output doesn't look any different. What happened? Well, one problem is that you're zoomed in too much to see the line at 3.14. Zoom out by scrolling up with your mouse wheel until you see a second line, as shown in Figure 3-12.

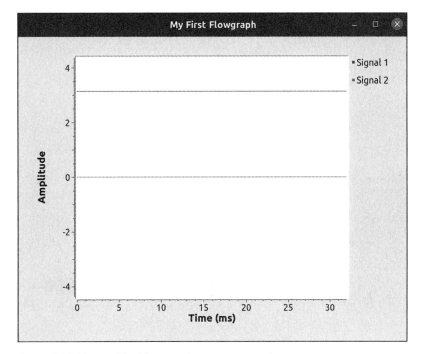

Figure 3-12: The modified flowgraph output, zoomed out

The output has changed as expected, with a constant line drawn at 3.14. Well, sort of. The line at 3.14, called Signal 1, makes sense, but what's this other line called Signal 2, and why is it at 0? The short answer is that you've just had your very first complex number sighting in GNU Radio. Unlike the typical numbers we deal with in everyday life, complex numbers have two parts, and you're seeing both of them here. We'll run into complex numbers again before long, but for now we don't need to dive into them any deeper.

Between Input and Output

You've seen how to get data into your flowgraph with source blocks and how to get it out with sink blocks. What do you do in between? The short answer is *math*! Most of the other blocks in GNU Radio Companion are devoted to performing mathematical operations to modulate, demodulate, filter, or otherwise manipulate signals as they move from input to output. To get a feel for how it works, update your flowgraph so that the constant signal is modified by a mathematical function.

Click **File ▶ Open** to open your *source_sink.grc* file (if you don't still have it open). Then click **File ▶ Save As** and resave it as *simple_multiply.grc*. Next, type **multiply** into your search window and double-click the `Multiply Const` block to add it to your flowgraph.

This block provides one of the simplest mathematical functions possible between your source and sink: it multiplies the incoming signal by a constant value and outputs the result. Before you can use it, however, you need to eliminate the connection that already exists between Throttle and QT GUI Time Sink. Right-click the line connecting the two blocks, and select **Delete** from the context menu. Alternatively, you can left-click the connection to select it and then press DELETE.

With the old connection out of the way, connect the Throttle output to the input of the Multiply Const and the Multiply Const output to the QT GUI Time Sink input, as shown in Figure 3-13. You can make the connections in any order.

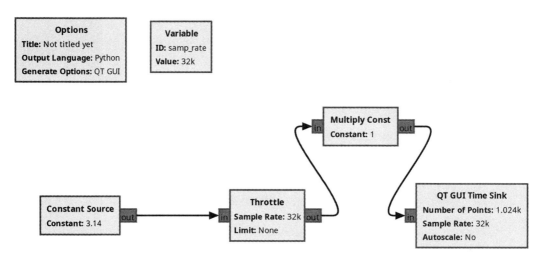

Figure 3-13: A simple flowgraph with multiplication

By default, Multiply Const multiplies the incoming data by a constant value of 1. Change the multiplier to 2 by double-clicking the block and updating the Constant property. The window should then look like Figure 3-14.

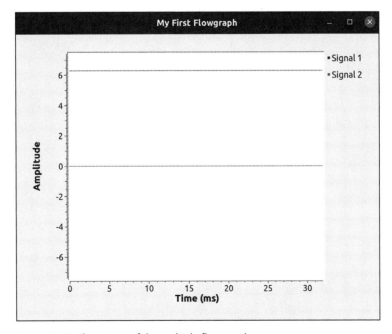

Figure 3-14: The `Multiply Const` window

After clicking **OK**, run the flowgraph again. You'll need to zoom out even more (remember to scroll up), but when you do, you'll see that the new value of Signal 1 is twice the previous one, as shown in Figure 3-15.

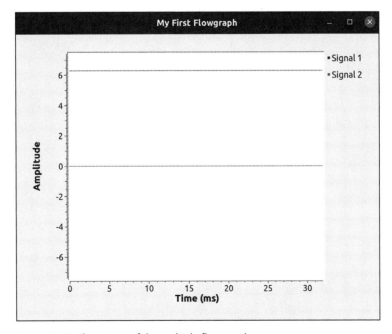

Figure 3-15: The output of the multiply flowgraph

Multiplying by two isn't exactly rocket science, but it illustrates an important point: between the input and output, all you're really doing is math. Before long, you'll be packing all sorts of interesting things between your sources and sinks.

Conclusion

At this point, you've got a very basic understanding of how to use GNU Radio Companion. This includes how to add blocks to a flowgraph as well as how to connect and configure them. You can delve much deeper into the capabilities of GNU Radio Companion, but you know enough right now to build your first radio. So let's do it!

4

CREATING AN AM RECEIVER

In this chapter, we'll dive right into building your first software-defined radio: an amplitude-modulated (AM) receiver. Like an AM car radio, it will be able to take in AM radio signals and convert them to listenable audio. Rather than work with live radio signals, however, we'll test out the radio with a file containing captured radio data. This way you won't have to worry about any hardware just yet.

At this early stage of your SDR journey, there are two main things you need to learn: how to use GNU Radio Companion and the theory behind how SDRs are built. This chapter will focus on the former. As such, for now, we'll mostly gloss over the radio theory behind the receiver we build. Without this theory, some of the steps you take may not make sense yet, and it may feel like you're following a rote set of instructions, almost like building a model airplane. Rest assured, though: the intention is not to give you simple cookbook instructions and send you on your way. We'll keep coming back to this project in the next few chapters to dig into the details of how the AM receiver, and radios in general, work.

Open up GNU Radio Companion and let's get started!

Setting Up the Variables and Entries

Create a new flowgraph by clicking **File ▸ New ▸ QT GUI** or by pressing CTRL-N. This will bring up a mostly empty starting flowgraph with just two blocks already in the workspace. Every new flowgraph starts with both of these blocks. The first is the Options block, which contains some basic documentation and settings for the flowgraph. The other is a Variable block, representing the variable called samp_rate. In general, you'll employ these Variable blocks to store values used throughout the flowgraph, just like defining variables in text-based programming languages like C or Python. As you might guess, this particular variable has something to do with the sample rate for the flowgraph.

Although not strictly necessary, it's good practice to add some basic information to the Options block. Double-click the block and then change the Title to **AM Receiver** and the Description to **My first AM radio receiver**.

Next, add a QT GUI Entry block to the flowgraph, as shown in Figure 4-1. (Remember, you can find new blocks using CTRL-F and the search box.)

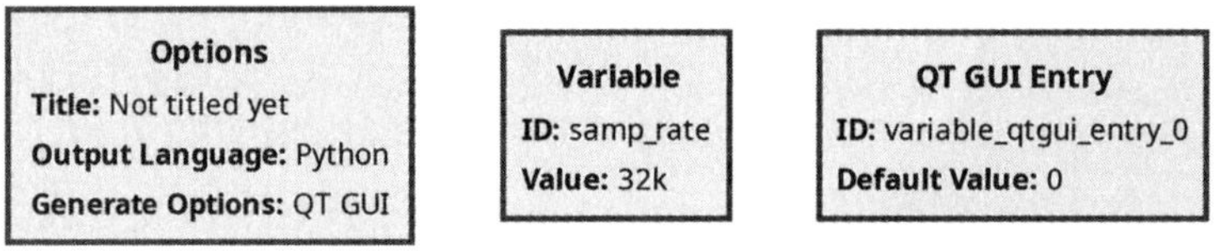

Options	**Variable**	**QT GUI Entry**
Title: Not titled yet	**ID:** samp_rate	**ID:** variable_qtgui_entry_0
Output Language: Python	**Value:** 32k	**Default Value:** 0
Generate Options: QT GUI		

Figure 4-1: Adding a QT GUI Entry

Like a Variable block, a QT GUI Entry stores a value for use throughout the flowgraph. Unlike a Variable block, however, you can change a QT GUI Entry block's value in real time while the flowgraph is running. We tend to put our Variable and QT GUI Entry blocks together at the top of the flowgraph, but this isn't necessary. The flowgraph will function the same regardless of the position of the blocks in your workspace.

Double-click the new **QT GUI Entry** block to bring up its properties window. Change the ID to freq and the Default Value to 880e3. This is exponential notation for 880,000. It's a common programming-language way of rendering 880×10^3. You can think of the value after the e as the number of zeros that GNU Radio Companion will add to the end of the number before the e. Exponential notation is especially useful for large numbers, making them much easier to read. For instance, 600000000 is not as clear at first glance as 600e6. Remember this notation. You'll be using it a lot.

There's just one problem with what you've typed: if you click the Apply button, you'll see an error at the bottom telling you that the Default Value is invalid, as shown in Figure 4-2.

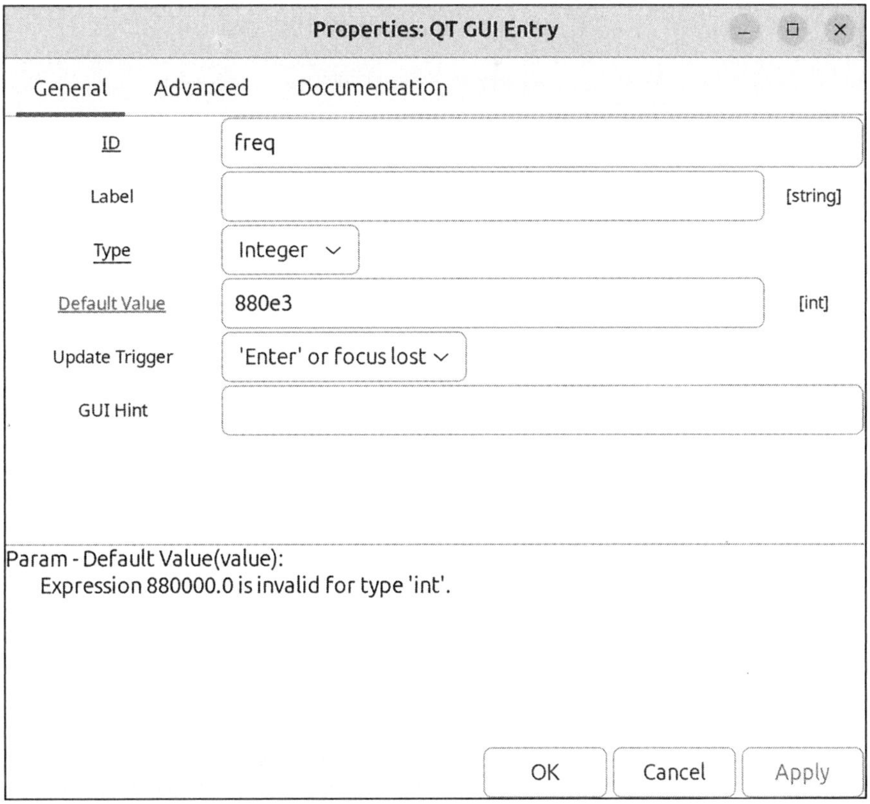

Figure 4-2: The QT GUI Entry error

The problem here is related to *data types*. You learned that flowgraphs can be thought of as numbers flowing out of sources, through blocks, and into sinks, but you haven't learned anything about what kind of numbers those are. Just as in programming languages like Python, Java, or C++, there are different types of data in GNU Radio Companion. As you can see on your screen (and in Figure 4-2), the Type property of the QT GUI Entry block is set to Integer, meaning it can accept positive or negative numbers without a decimal point, like 17 or -1293. It turns out, however, that exponential notation produces a floating point–typed value: a number with some digits after the decimal point, such as 3.14159 or -8.9. To make the Type of the QT GUI Entry compatible with the Default Value we've given it, click the pull-down menu containing Integer and instead select **Float**. Then click **Apply**, and the error should go away, as shown in Figure 4-3.

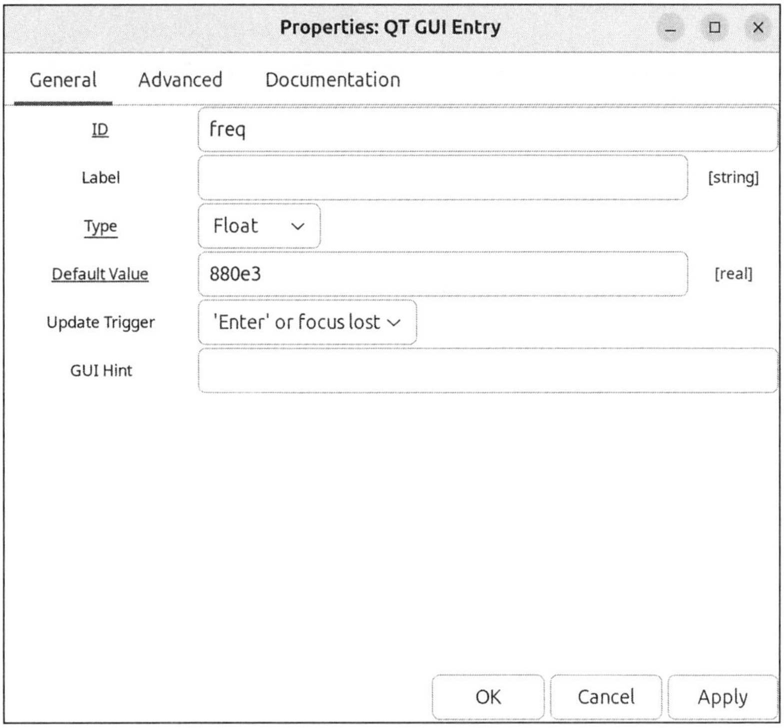

Figure 4-3: The corrected QT GUI Entry block

Click **OK** and note that the rendering of the block in the workspace changes to reflect the value you've entered. This is useful because even as our radio flowgraph gets progressively more complicated, you can see a lot about how it works from a top-level view, all at a glance.

Next, you'll add a second QT GUI Entry with a different ID and Default Value. Instead of adding it like before, though, you can simply copy and paste the first one by clicking the existing **QT GUI Entry** and pressing CTRL-C (or selecting **Edit ▸ Copy** from the menu bar), followed by CTRL-V (or **Edit ▸ Paste**). Notice that the new block appears with freq_0 as its ID, as shown in Figure 4-4.

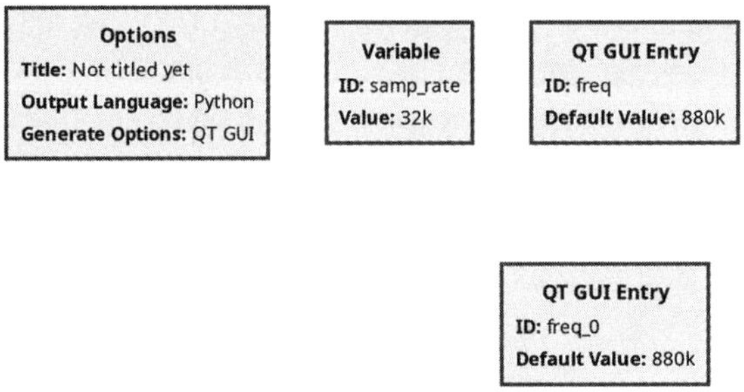

Figure 4-4: Copying the QT GUI Entry block

Double-click the new block and set its ID to center_freq with a Default Value of 900e3. Because you copied the block, it already has its Type set to Float. When you're done, your flowgraph should look like Figure 4-5.

Figure 4-5: The second QT GUI Entry, configured

Notice that we've moved the new QT GUI Entry up to the top with the other blocks. Again, you may find it useful to keep all the Variable and QT GUI Entry blocks together at the top of the flowgraph.

Adding a Source of Radio Data

Next, we'll add a source block to bring radio data into the flowgraph. First, grab a File Source block and add it to the workspace. This will allow you to use a file as the radio data input to the flowgraph, so you won't require any SDR hardware. The file contains actual raw radio data that an enterprising SDR aficionado captured from the airwaves and stored for later use. If you were building a fully working radio, you'd use a different source block instead of the File Source. This alternative source would interface with your SDR hardware and provide real-time radio data to your flowgraph.

As before, double-click the newly placed block to set up its properties. In the File selection, click the three dots, then navigate to the location of the project files you downloaded earlier from *https://nostarch.com/practical-sdr*. Select the one named *ch_04/am_ broadcast_02_c900k_s400k.iq*. You'll see a warning in the lower portion of the properties window telling you that a port isn't connected, but you'll fix that in a moment, so don't worry about it. At this point your property window should look similar to Figure 4-6. Click **OK** to return to the workspace.

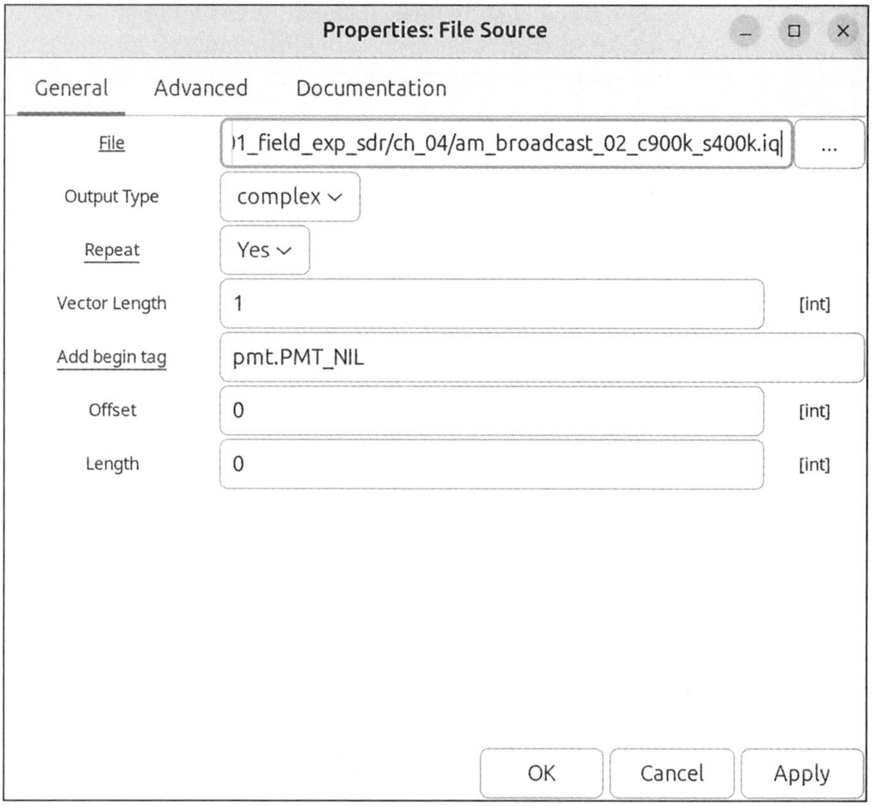

Properties: File Source

General   Advanced   Documentation

File	)1_field_exp_sdr/ch_04/am_broadcast_02_c900k_s400k.iq	...
Output Type	complex ∨
Repeat	Yes ∨
Vector Length	1	[int]
Add begin tag	pmt.PMT_NIL
Offset	0	[int]
Length	0	[int]

OK   Cancel   Apply

Figure 4-6: The File Source *block properties*

Now's a good time to save the project, so press CTRL-S or select **File ▸ Save**. We can give it any legal Linux filename, but let's use *first_am_rx.grc* (*rx* is shorthand for *receiver*). When working on your flowgraphs, remember to save early and often.

NOTE   *From now on, we're going to assume you know how to search for blocks and add them to the design. We'll also assume you know how to bring up the properties listing for a block and change the relevant values. As such, we'll stop spelling out each step of these processes.*

Processing the Signals

The next several blocks we'll add will work together to process the radio signal from the File Source. Again, in this chapter we won't focus much on the details of what these blocks are doing or how exactly they work; we'll explore those questions later in the book. For now, our concern is building a working flowgraph.

First, place a Signal Source block into your workspace. You'll use this to generate an infinitely repeating sequence of values representing a sinusoid.

In the block's properties, you're going to do something a little different: for the Frequency property, you aren't going to enter a simple number. Instead, type center_freq - freq, as shown in Figure 4-7.

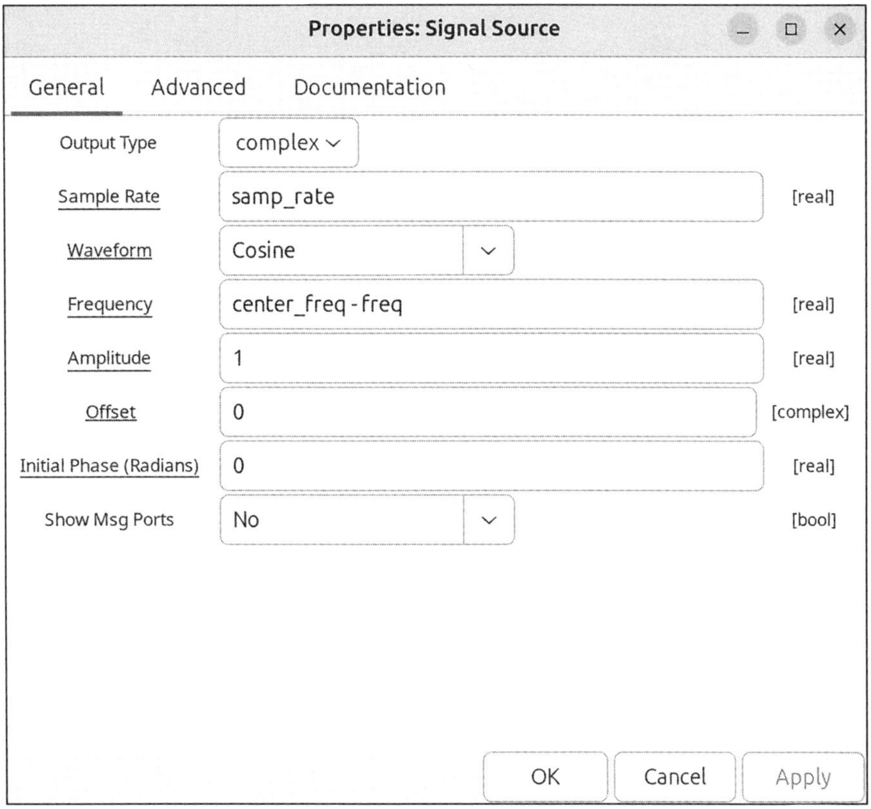

Figure 4-7: The Signal Source properties

Can you see what you just did here? Instead of entering a fixed number for a property, you can enter variables. And not just variables, but mathematical expressions involving multiple variables. In fact, as you'll see later, you can enter almost any legal Python expression as a property for a block and it will work. This will turn out to be very useful.

In this case, you're setting the frequency of the Signal Source using the two QT GUI Entry blocks you created. Specifically, you're subtracting the value of the block with ID freq from the value of the block with ID center_freq. Notice that when you click **OK** and go back to your workspace view, you don't see the math expression you just typed, but simply the number that results from it, as shown in Figure 4-8.

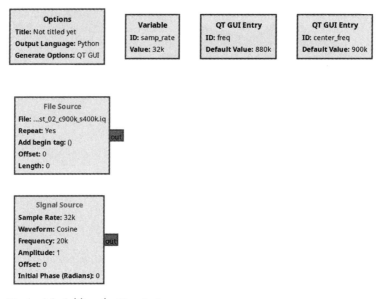

Figure 4-8: Adding the Signal Source

In the workspace, the Signal Source frequency is listed as 20k, or 20,000. This is equal to the center_freq default value of 900e3 (900,000) minus the freq default value of 880e3 (880,000).

Next, place a Multiply block in the workspace and connect its inputs to your two sources (make sure you don't grab the wrong block, since there are a lot with the word *Multiply* in them). Remember how to connect blocks? Simply click the output port of the first block (in this case, one of the sources) and then click an input port of the second (in this case, the Multiply block). Clicking in reverse order also works. But which ports are the Multiply inputs? If you look closely at the text inside the tabs, you'll see that two of them say in0 and in1. That's them! In general, the inputs will be on the left and the outputs on the right, but sometimes blocks will be rotated and this will no longer be true. When you're done, the flowgraph should look like Figure 4-9.

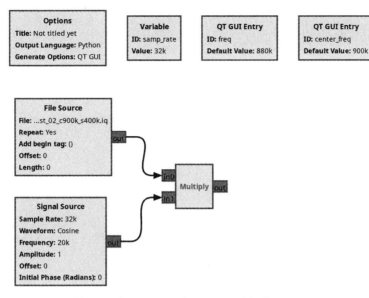

Figure 4-9: Adding and connecting the Multiply *block*

As you add and connect more blocks, it's good to take a moment here and there to tweak the positioning to tidy up your flowgraph. Just click and drag the blocks where you want them to go. As you do so, notice that the connections you've made are sticky and will follow your blocks wherever you drag them. This neatening process isn't required but can make your flowgraph much easier to read.

Next, add a Low Pass Filter block, setting the Cutoff Freq to 5e3 and the Transition Width to 1e3, as shown in Figure 4-10. Leave all the rest of the properties alone.

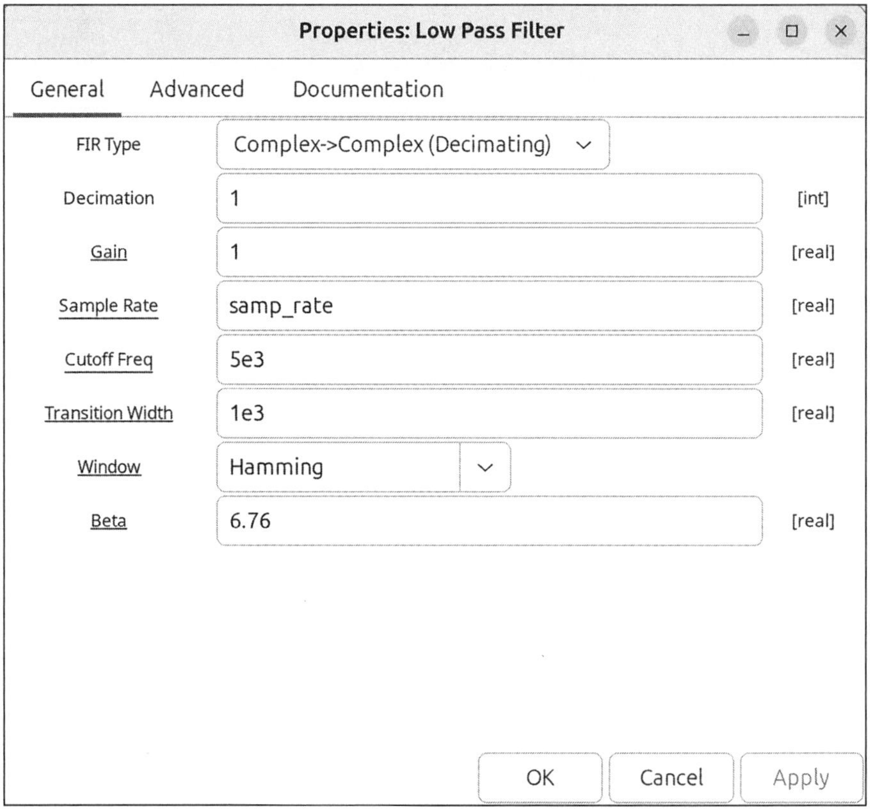

Properties: Low Pass Filter

General	Advanced	Documentation

FIR Type	Complex->Complex (Decimating) ⌄	
Decimation	1	[int]
Gain	1	[real]
Sample Rate	samp_rate	[real]
Cutoff Freq	5e3	[real]
Transition Width	1e3	[real]
Window	Hamming ⌄	
Beta	6.76	[real]

OK   Cancel   Apply

Figure 4-10: The `Low Pass Filter` properties

Filters like this block remove or reduce certain frequencies from a signal. They're an extremely important concept that we'll explore in detail in Chapter 5. Connect the `Low Pass Filter` input to the `Multiply` output. You should now have a workspace similar to Figure 4-11.

So far, here's what's happening in the flowgraph: the `File Source` block brings in data, which is then processed using the other three blocks (the `Signal Source`, `Multiply`, and `Low Pass Filter`). As we'll discuss in Chapter 6, these three blocks comprise the AM radio's tuner. They're what will allow you to focus in on individual radio channels within the source data.

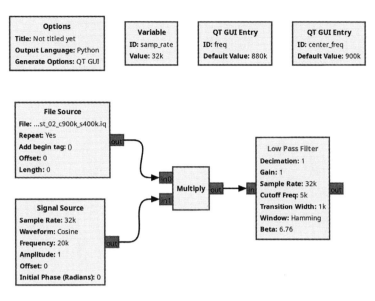

Figure 4-11: Adding the `Low Pass Filter` *block*

At this stage, the `Low Pass Filter` block's label should be red (not shown in Figure 4-11, as the book is grayscale), while the other block labels are black. This happens because GNU Radio Companion actively checks your flowgraph for errors as you build it, and any blocks with illegal conditions have their title displayed in red text. The illegal condition in this case is that the `Low Pass Filter` block's output isn't hooked up to anything. Other common illegal conditions include missing or invalid properties or multiple outputs connected to the same input.

Now place an `AM Demod` block and set its Channel Rate to `samp_rate` and its Audio Decimation to 1, as shown in Figure 4-12.

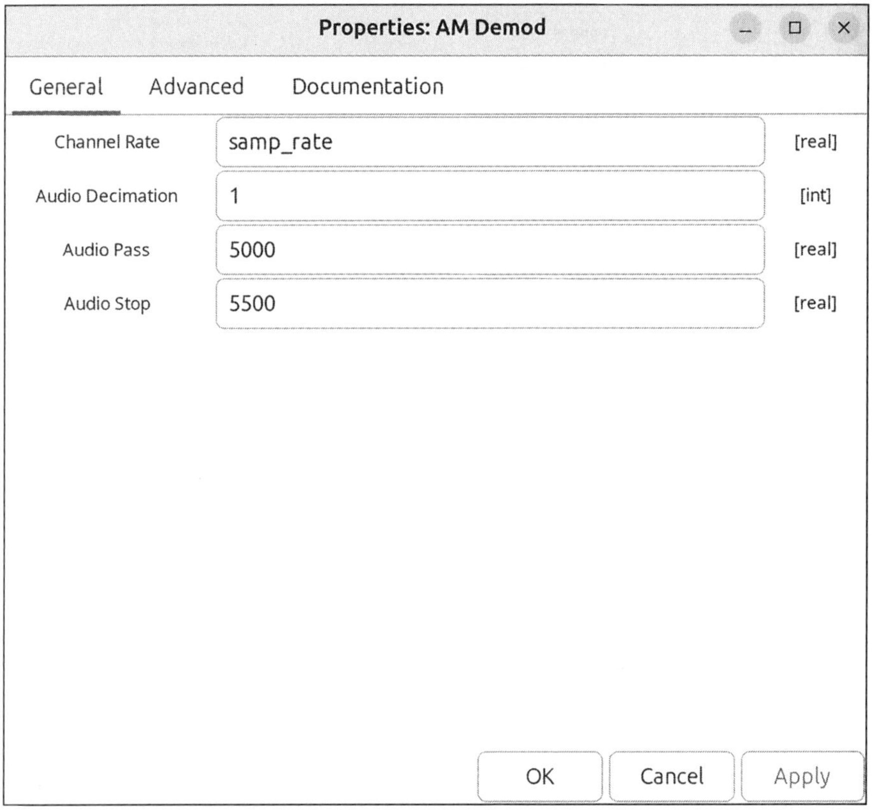

Properties: AM Demod		
General Advanced Documentation		
Channel Rate	samp_rate	[real]
Audio Decimation	1	[int]
Audio Pass	5000	[real]
Audio Stop	5500	[real]

OK Cancel Apply

Figure 4-12: The AM Demod block properties

The AM Demod block will demodulate the incoming radio signal, extracting a human-understandable audio signal from it. We'll explore how its settings work in Chapter 6. Connect its input to the Low Pass Filter output, as shown in Figure 4-13.

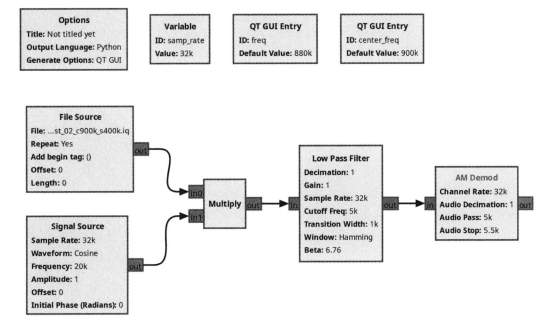

Figure 4-13: Adding the AM Demod block

If you're following along on your computer, notice that the input port of the AM Demod block is blue, while its output port is orange. This is significant: the colors represent the type of data that flows into or out of the block. Orange ports mean floating-point numbers are flowing, while blue ports mean complex numbers. This is definitely *not* the right time to get into complex numbers, so for now, just be aware that connections can only be made between ports of the same color.

Next, place a Rational Resampler block and set its Interpolation to 32 and the Decimation to 400, as shown in Figure 4-14.

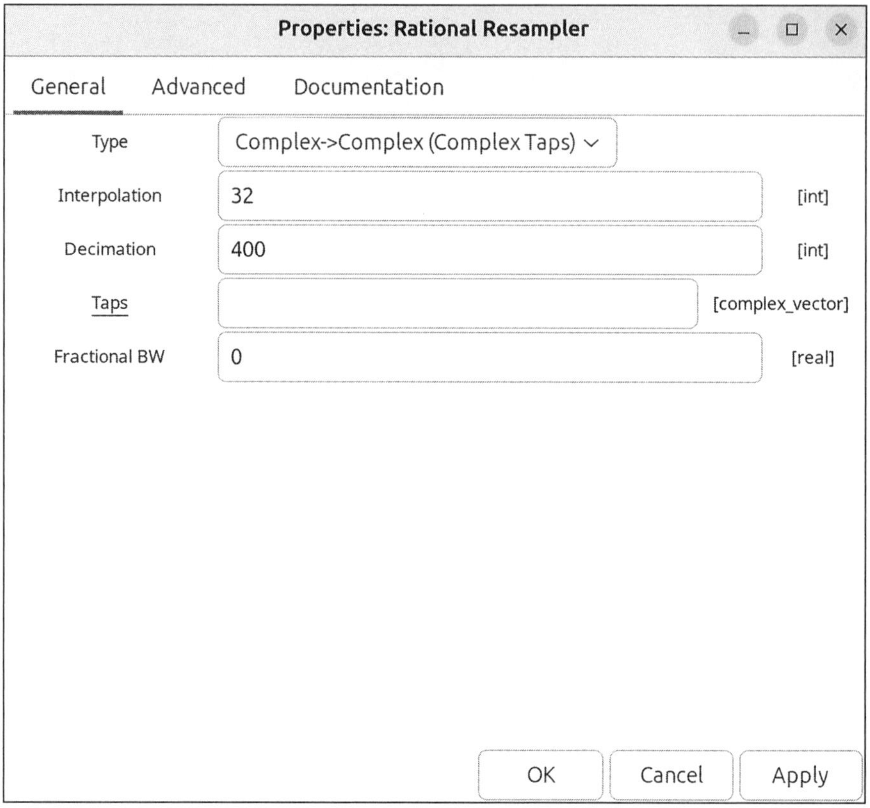

Figure 4-14: The `Rational Resampler` *properties*

The `Rational Resampler` block will adjust the sample rate of the audio signal so that your sound card can play it. Connect its input to the `AM Demod` output, like you see in Figure 4-15.

But not so fast! The connection should appear on your screen as a red arrow. Something is wrong, and there was a hint as to what a couple of paragraphs ago.

You can't connect ports of different colors. Thinking again in terms of programming languages, this would be like passing a string parameter to a function that's expecting an integer. To fix the problem, open up the `Rational Resampler` properties again and note the Type property is currently `Complex->Complex (Complex Taps)`. Instead, select **Float->Float (Real Taps)**, then click **OK**. You should now see that the `Rational Resampler` block's ports are orange, and the connection from the `AM Demod` to the `Rational Resampler` should have changed from red to black.

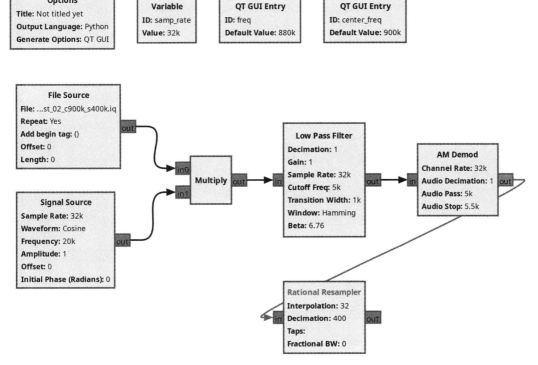

Figure 4-15: Adding the Rational Resampler *block*

We're almost done with the flowgraph. Take a moment to think about what might be missing. You have a source (two of them actually), and you have some blocks that process the data coming in from that source. But what are you going to do with your processed data stream? You need to dump it into a sink!

The Output

To output the data from the flowgraph, select an **Audio Sink** and add it to your workspace. This block will "play" the data through your computer's sound card so that you can hear the sounds being broadcast. Double-click the block and select a value of **32kHz** from the pull-down menu for the Sample Rate property. Then connect the output of the Rational Resampler to the input of the Audio Sink. With a bit of flowgraph tweaking, you could select a different sample rate, but most computer audio hardware supports the 32 kHz rate.

There's one last thing to do: change the sample rate for the flowgraph. Did you notice how most of the blocks have a Sample Rate of 32k displayed? This is because all the blocks had a default Sample Rate property equal to samp_rate. If you remember, a `Variable` block was present when you started the project. It had an ID of samp_rate and a value of 32000. Double-click this variable and change its value to 400e3, then watch what happens. You can also see the result in Figure 4-16.

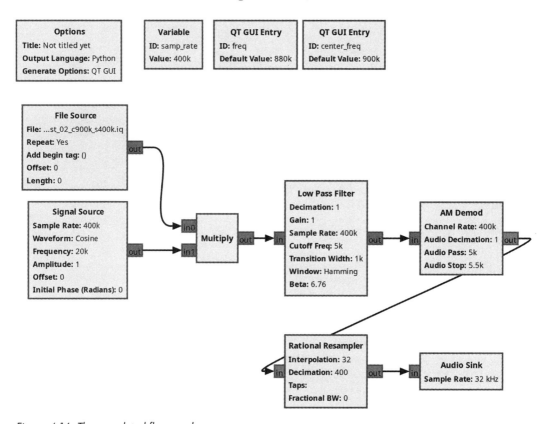

Figure 4-16: The completed flowgraph

Changing a single `Variable` block caused changes to ripple throughout the design such that all the blocks now have a sample rate of 400k. All but the last one, anyway.

Run the flowgraph to see if it works. As in the last chapter, just click the toolbar icon that looks like a play button. If you hover over the button, it will say "Execute the flowgraph."

When you execute, some text will start scrolling by in the console window pane. Then, after a few seconds, you should hear some music. There's a bit of static, as you may have heard on other AM radios, but there are clearly identifiable voices. The music will loop after a few seconds because there's not a lot of data in the `File Source`, and it's set to repeat. If you don't hear any audio, try adjusting the speakers on your computer.

If you experience choppy audio and are running GNU Radio Companion in a virtual machine, open the properties window of the Audio Sink and change the Device Name to sysdefault, as you can see in Figure 4-17. If you aren't working in a VM, this isn't advised.

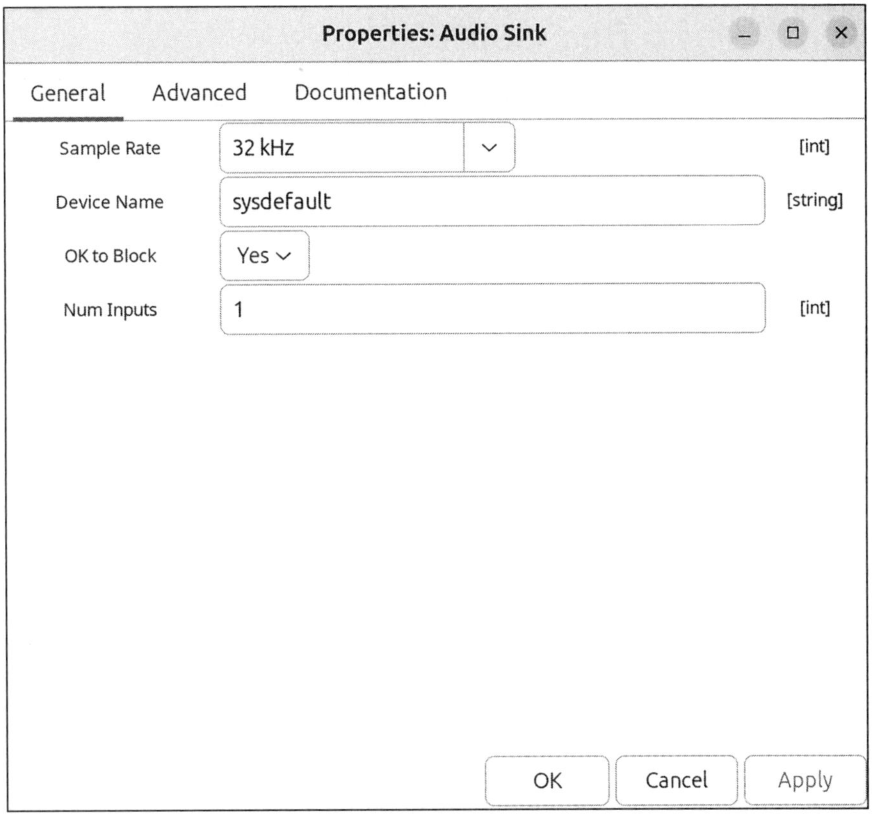

Figure 4-17: The Audio Sink setup for virtual machines

As the flowgraph runs, a window will pop up with the two QT GUI Entry elements in it, as shown in Figure 4-18. This is the *execution window.*

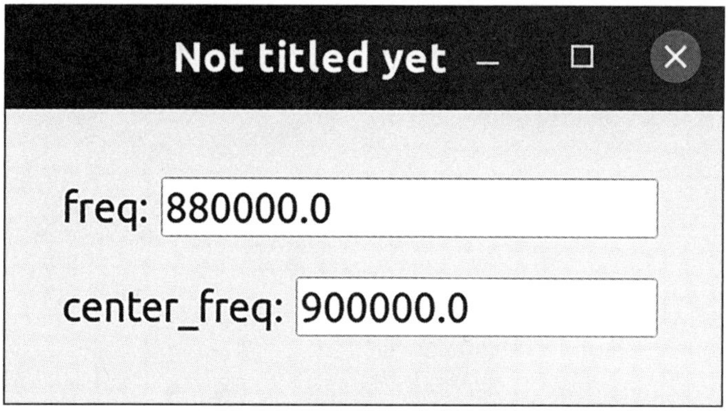

Figure 4-18: The flowgraph execution window

As mentioned early in this project, you can change these QT GUI Entry values while the radio is running, whereas to change an ordinary Variable block, like samp_rate, you'll have to stop the flowgraph, change the value, and run it again. In the execution window, change the value of freq to 750k, then press ENTER. After a moment, you'll hear a different repeating chunk of audio. That's because you've now tuned to a different radio station! You can also find stations at 710k, 750k, and 1000k, as well as a few fainter ones at 950k and 1090k.

Take a moment to think about what you just experienced. The person who created the input file you used didn't just record the *audio* from an AM station. That's pretty easy to do. They recorded the raw *radio signal* from that station, from which your flowgraph extracted recognizable audio. What's more, they didn't just record the radio signal broadcast on a *single* AM channel. They recorded the signals on a whole bunch of channels so that you can tune to any of them whenever you want.

This input file (and your flowgraph) provides your first glimpse of an extraordinary SDR capability: the ability to capture raw radio data and process it at will.

Conclusion

In this chapter, you built a simple AM radio receiver in GNU Radio Companion. Without going into a lot of detail yet, let's peel back one layer of the onion on your radio design. At a very high level, here's what your AM radio receiver flowgraph is doing:

1. Injecting prerecorded radio data into the flowgraph
2. Tuning to a specific AM radio channel while filtering out other AM radio channels
3. Demodulating the signal of your desired channel
4. Doing some magical thing called *resampling*
5. Playing the resulting audio on your computer speakers

The blocks responsible for each of these five tasks are highlighted in Figure 4-19.

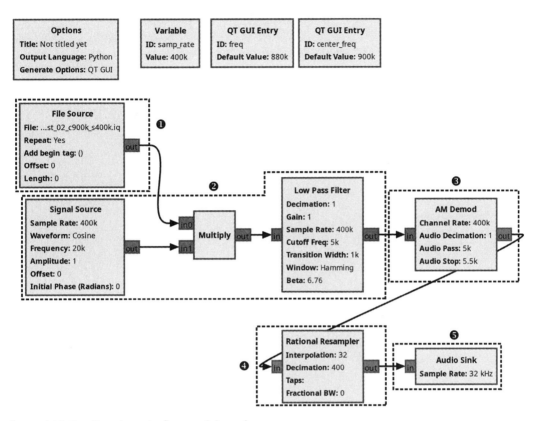

Figure 4-19: Breaking down the flowgraph by task

The File Source brings the data into the flowgraph ❶. The Signal Source, Multiply, and Low Pass Filter blocks work together to tune the signal ❷, which is then demodulated as it passes through the AM Demod block ❸. Then the signal is resampled by the Rational Resampler ❹ before being output by the Audio Sink ❺.

You probably have a few questions. How does a filter work? What does it even mean to resample? How on earth does multiplying a radio signal by a sinusoid tune anything? You'll get the answers these questions in the next few chapters, partly so you can understand how your AM radio flowgraph works, but even more so because the answers will illuminate some critical radio concepts.

PART II

INSIDE THE RECEIVER

5

SIGNAL PROCESSING FUNDAMENTALS

Now that you know your way around GNU Radio, this chapter will explore some of the basic radio and signal theory underpinning the AM receiver you built in Chapter 4. We'll discuss the concepts of *frequency* and *gain* and then see how these two concepts come together in one of the most commonly used signal processing tools: *filters*. You may have noticed that your AM receiver from the last chapter contained a filter block, and this is not at all unusual. As you move forward learning to build different types of radios with GNU Radio, you'll find that nearly all of them will contain at least one filter, so it's important to understand how filtering works.

Frequency

Physicists define *frequency* as the number of oscillations of a periodic phenomenon per unit time. You can think of a periodic phenomenon as any event that repeats over and over with a consistent timing. Frequency is typically measured in hertz (Hz), where 1 Hz is equal to one cycle per second. Notably, this definition of hertz doesn't specify the thing doing the cycling. This is because frequency is a property of all sorts of phenomena. For example, you may associate frequency with stations on your radio dial, the specifications of your wireless router or mobile phone, or perhaps the speed of your computer processor, probably given as some number of gigahertz (GHz) or megahertz (MHz).

From the standpoint of radio theory, we're most interested in the frequencies of radio waves, but there's another useful occurrence of frequency that you probably already have a pretty good instinctual handle on: sound. In this section, we'll explore the concept of frequency through the more intuitive lens of sound waves. As we do, keep in mind that much of what you'll learn applies to radio waves as well.

Exploring the Audible Spectrum

The typical human ear can detect sounds roughly between 20 Hz and 20 kHz (depending on age and the number of heavy metal concerts attended). We call this range of frequencies the *audible spectrum.* The frequency we're considering here represents how many times per second the air vibrates when a particular sound is present. Higher-frequency air vibrations than 20 kHz exist, but we call them *ultrasonic.* Your dog can hear some of these frequencies, but you can't. Lower-frequency air vibrations than 20 Hz exist as well, but we call them *subsonic* and likewise can't hear them. Figure 5-1 summarizes the whole range of sound wave frequencies.

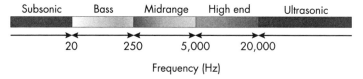

Figure 5-1: Audio frequencies, including the audible spectrum

Between the subsonic and ultrasonic extremes, the audible spectrum of frequencies encompasses a wide variety of sounds. In the bass range, there are low-pitched (low-frequency) sounds that are like a low rumble. At the high end, there are high-frequency sounds that resemble piercing shrieks. And there's the midrange in between.

Generating a Tone

We'll begin a hands-on exploration of frequency and the audible spectrum by generating a simple audio tone. Conveniently, we can use GNU Radio to work with audio frequencies, just as we can with radio frequencies. The software sees the audio data as nothing more than a bunch of numbers and has no way of knowing that the numbers didn't come from a radio, so let's go back to GNU Radio Companion and pretend for a moment that it's actually GNU *Audio* Companion.

Start a new flowgraph and place a Signal Source block, then set the Output Type to Float, the Frequency to 500, and the Amplitude to 0.1. This block will generate a stream of floating-point numbers that can be interpreted as a simple 500 Hz audio signal. In general, any audio signal, whether a real-world signal generated by a microphone's output voltage or a synthetic one like we're using here, can be represented by a stream of floating-point numbers. This is why we'll be using the Float type for the audio waveforms in this chapter, in contrast to the Complex type we'll use elsewhere for radio signals.

Add an Audio Sink and connect it to the Signal Source so that you can hear the signal. Then save this flowgraph as *single_tone.grc*. When you're done, you should have something like Figure 5-2.

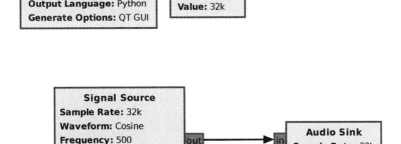

Figure 5-2: The single-tone flowgraph

When you execute this flowgraph, you should be able to hear a single sustained note. This is the sound of a 500 Hz tone. Looking back at Figure 5-1's frequency range diagram, 500 Hz falls within the midrange. Listening to the sound of the tone, this should be unsurprising: it isn't especially low pitched, nor is it particularly high pitched.

NOTE   *If you can't hear anything when you execute the flowgraph, try using headphones. Also, if you're running on a virtual machine, try setting the Device Name property of the Audio Sink to sysdefault.*

Visualizing the Tone

You've heard what a 500 Hz tone sounds like, but what does its waveform look like? To find out, add a QT GUI Time Sink block to the flowgraph, set its Type to Float, and connect it to the Signal Source output, as shown in Figure 5-3.

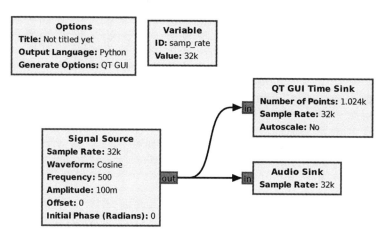

Figure 5-3: A simple tone with a QT GUI Time Sink block

Notice that GNU Radio Companion lets you attach the same output to multiple block inputs. In this case, the Signal Source output goes to both the QT GUI Time Sink and the Audio Sink to provide both audio and visual representations of the signal. It's a good habit to add different types of monitors and displays throughout your flowgraph so you can be sure that you're getting the signals you expect at each point.

When you execute the flowgraph again, along with hearing the tone, you should see a window appear with a repeating, curving shape, as shown in Figure 5-4. This is what the output of our Signal Source looks like, plotted against a time axis.

What you have here is a *sinusoid*, much like the ones we discussed in Chapter 1, gradually alternating between −0.1 and +0.1. Any pure, single-frequency tone, such as this one at 500 Hz, is sinusoidal in nature. The explanation why is long and mathematical. For now, you'll have to take it on faith that sine waves produce pure tones.

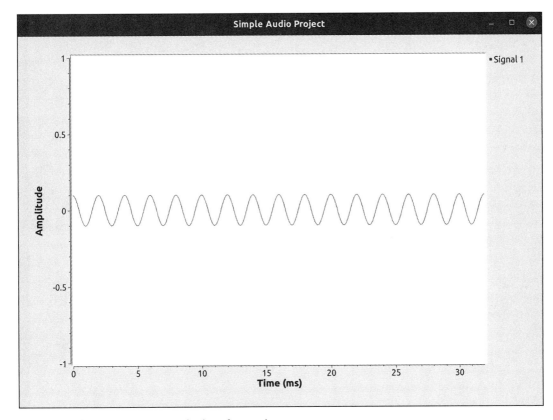

Figure 5-4: The QT GUI Time Sink display of a simple tone

You can see only about 30 milliseconds (ms), or 0.03 seconds, of the signal in the plot, but based on the way it's repeating, there's no reason to expect that displaying a larger period of time would show anything different. To get a closer look at the signal, you can zoom into the waveform in two ways. Rolling the scroll wheel of your mouse up and down will zoom in and out vertically. You might remember doing this in your first project in Chapter 3. You can also zoom by left-clicking your mouse anywhere on the display and holding the button while drawing a box. When you release the mouse button, the window will change to display the contents of the box you drew. Try it now: zoom in on a single oscillation, or sequence of "up" and "down," to get the image you see in Figure 5-5.

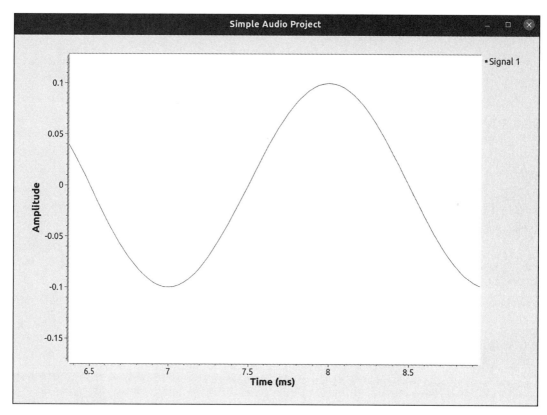

Figure 5-5: Zooming in on the simple tone

To zoom back out, simply click the right mouse button. You can actually zoom in multiple times by drawing a zoom box and then drawing another zoom box inside of the zoomed display. The right mouse button will just undo the most recent zoom operation. This nested zooming can be useful for complex signals.

Varying the Tone's Frequency

We'll now try varying the frequency of the generated tone to get an idea of what different parts of the audible frequency range sound like. It would be easiest if you could adjust the frequency of the tone while the flowgraph is running rather than stopping, editing, and restarting the flowgraph with each change. In Chapter 4 we used the QT GUI Entry to provide this functionality, but this time we'll use a different block with a handier user interface. Add a QT GUI Range to the flowgraph, changing the ID to freq, the Default Value to 500, the Start to 100, the Stop to 32000, and the Step to 100. This block will let you control the value of freq graphically using a slider. The extra properties (Start, Stop, Step) define the range and granularity of the slider. Next, in the Signal Source block, change the Frequency to freq. When you're done, your workspace should look like Figure 5-6.

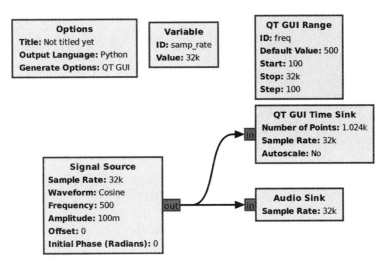

Figure 5-6: A simple tone set using a QT GUI Range block

When the Signal Source generates a signal, it will now look to the QT GUI Range for the signal's frequency, which you'll be able to vary in real time. Rerun the flowgraph, and you'll have a single window containing the time display and a slider for the frequency control, as shown in Figure 5-7.

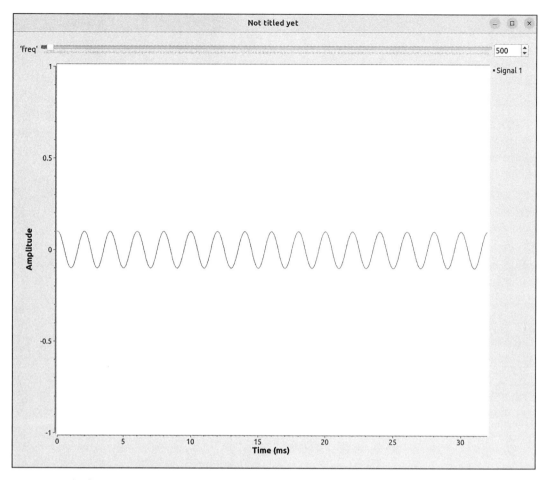

Figure 5-7: The flowgraph execution window with frequency control

Try changing the value of freq by clicking and dragging the slider left and right. The sound of the generated tone should vary accordingly. As the frequency value changes, it might be necessary to zoom in or out to get a better look at the waveform. Also, keep in mind that both your sound card and your ears are imperfect, so you may not hear a tone when the frequency is very low or above 10,000 Hz.

Does the behavior of your flowgraph make sense? Depending on how thoroughly you played around with the frequency, the answer may be "Sort of?" If you have a very expensive sound card and exceptionally good hearing, the behavior should make sense up to 16,000 Hz: as the frequency increases, the tone gets higher and the sinusoid oscillates faster, going up and down more times in the same time period. If you keep going higher, however, something strange will happen: the tone will get progressively lower in pitch the higher you raise the frequency. What's going on here?

One clue is to look at the sample rate of the blocks. Remember that in Chapter 2 we said the sample rate needs to be fast enough relative to the signals being sampled. In this case, all the blocks in the flowgraph are sampled at 32 kHz, and the funny business starts when the sinusoid is approximately half that value. We won't do the math yet, but now we have some evidence that sampling twice as fast as the signals in the flowgraph is a good idea. *At least* twice as fast.

At this point you might be thinking, "So what?" We've made some annoying sounds and looked at some squiggly lines. What does this have to do with radios? Well, as you'll see in a moment, quite a bit. The frequency of a signal is a crucial property that we can use to identify, characterize, isolate, and process that signal. It doesn't matter whether you're working with audio signals, radio signals, or any other kind of signal; they all have frequencies, and they can all be treated the same way. To better understand how to use a signal's frequency to interact with that signal, we need to talk about an associated and very important concept in signal theory: welcome to the frequency domain.

Visualizing Signals in the Frequency Domain

So far we've been graphing audio signals in the *time domain*, watching how the waveform fluctuates over time, but it's also possible to graph an audio signal in the *frequency domain*. In this new and different type of visualization, frequency, not time, is depicted on the x-axis, and the graph indicates which frequencies are present in a given signal. Examining the frequency domain reveals that most sounds, with the exception of pure tones, are made up of a whole bunch of different frequencies that are simultaneously in play. The same is true for radio signals.

If you think about it, most of the sounds you encounter in normal life don't seem at all like the pure tones you just generated. They sound fundamentally different, and yet there's still this idea of sounds that are low, like a thumping subwoofer, and sounds that are high, some so high you can't even hear them (imagine a dog whistle). Also, consider that some sounds are low and high at the same time. When you're listening to music, for example, you can often hear the lower-pitched bass portions at the same time as mid-range singing and higher-pitched instruments.

If that rumbly bass sound coming out of the subwoofer isn't a pure tone like we've been making, then it must be something else, and yet it still must have certain frequency characteristics in order to sound low. If you examined the bass sound in the time domain, you'd find that it has a more intricately shaped waveform than the simple sinusoid of a pure tone. Examine it in the frequency domain, and you'd find that the sound contains multiple predominantly low frequencies, not just a single one. These multiple frequencies are what set the rumbly sound apart from a pure tone, and the fact that the frequencies are predominantly low is what makes the rumble seem low to us.

Viewing a sound in the frequency domain requires a mathematical tool that can break the sound down into its constituent frequencies. This mathematical tool is called the *Fourier transform*, and crucially it can be applied not just to sounds but to any kind of signal, including radio signals. We're too high up on the outer skin of the onion to talk about the mathematics behind the brainchild of Joseph Fourier, but here's a general explanation: we take a numerical representation of a signal, which you'll recall is just some value changing over time, and apply a mathematical operation to the signal to create a breakdown of all the frequencies it contains. Understanding how this works is central to understanding how SDRs can take in the whole mess of radio information bouncing around in the air and extract the particular signal you want.

Strictly speaking, Fourier transforms are for continuous signals; remember the signals with smooth curves? There's a counterpart called the *discrete Fourier transform (DFT)* that can do the same thing for sampled data, or the kind of digital data used in the SDR world. Because there are several ways to do the math for the DFT, we'll further specify the most common algorithm: the *fast Fourier transform (FFT)*. Viewing the FFT of real-world radio data is a common use of an SDR. SDR systems accomplish this by taking in a continuous radio signal, sampling it, running it through an FFT, and finding out which frequencies are present in that signal and how prominent each frequency is.

Viewing Simple Tones

Let's build a flowgraph to explore the frequency domain. Start with your last project (*single_tone.grc*), and resave it as *single_tone_fft.grc* (click **Save As** and provide the new filename). Then add a QT GUI Frequency Sink block and connect it to the Signal Source. This block will graph the output signal in the frequency domain, using an FFT to show what frequencies are present in the signal. The connection's red arrow is a warning that the input and output types don't match. After changing the QT GUI Frequency Sink block's Type parameter to Float and the Spectrum Width to Half, the flowgraph should look like Figure 5-8.

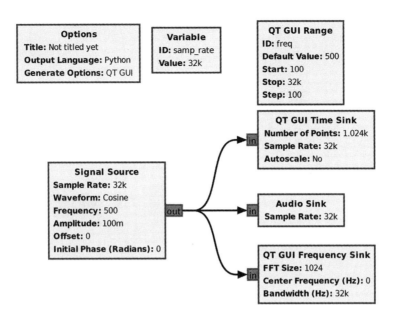

Figure 5-8: A simple tone with an added frequency display

Execute the flowgraph, and you'll see a frequency plot in addition to the time plot and frequency-adjustment slider. We're now viewing the signal in both the time domain and the frequency domain. At first, you'll see a single spike fairly close to the left side of the frequency plot, as shown in Figure 5-9.

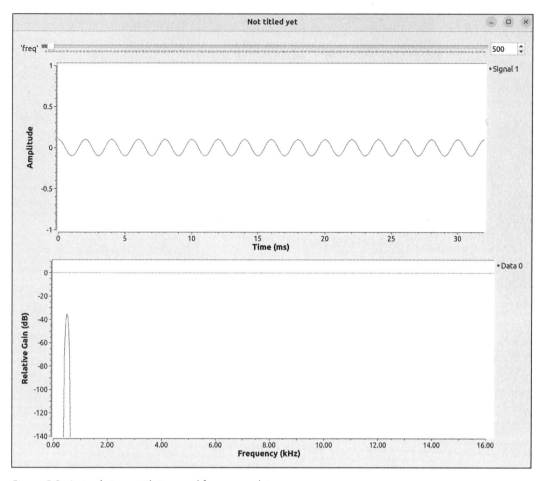

Figure 5-9: A simple tone with time and frequency plots

If you click and drag a box around the frequency plot spike to zoom in, you'll see that it resides at 0.5 kHz. This is shown in Figure 5-10.

Another way to express 0.5 kHz is 500 Hz, the default frequency set for the Signal Source, so it makes sense we're seeing a spike there. This is experimental validation of the earlier assertion that sinusoids produce pure tones with a single frequency. Is there any signal at 1,000 Hz or 100 Hz? No, the frequency plot shows nothing at those (and all other) values.

One complication is that the peak in the frequency plot isn't perfectly sharp. You could say that there's signal not just at 500 Hz but also some between 400 and 600 Hz. Even though the plot peaks at 500, the signal doesn't disappear until you get out of that 400 to 600 Hz range. This isn't because there are actually multiple frequencies present in the tone. Instead, it's because the FFT algorithm generating the frequency plot isn't mathematically perfect, but rather an approximation. Don't believe it? Let's prove it!

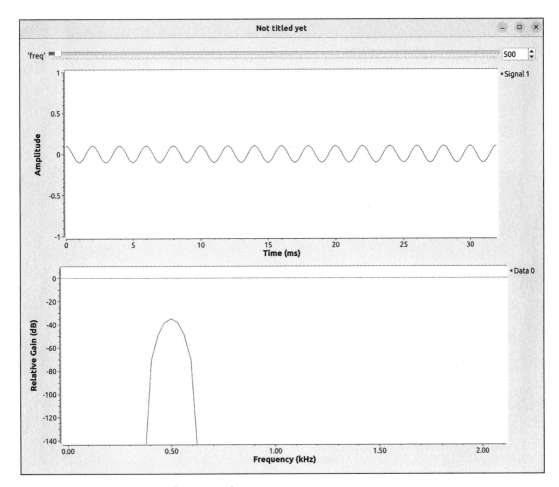

Figure 5-10: Zooming in on the frequency plot

Use the middle mouse button (scroll wheel) to click anywhere on the frequency plot. A context menu will pop up. Select **Control Panel** to bring up some extra control options for your frequency plot. Click the drop-down menu under FFT, and you should see numbers ranging from 32 up to 32768. Choose **4096**. The plot may automatically zoom back out, so go ahead and zoom back in, as shown in Figure 5-11.

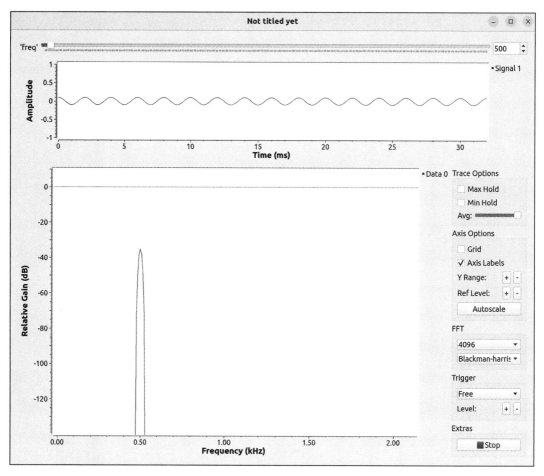

Figure 5-11: The frequency plot with an FFT size of 4096

Increasing the FFT size has yielded a narrower frequency spike; it now goes between about 470 and 530 Hz. The frequency plot is therefore more accurate, but this comes at the cost of requiring your CPU to do more computations. Feel free to experiment with different FFT sizes to see what happens.

Plotting More Complex Sounds

Now let's look at a sound with a bit more complexity: a chord containing three different tones. We'll see how this sound differs from a single pure tone in both the time domain and the frequency domain. Save your existing flowgraph as *cmajor.grc*. Next, modify your flowgraph to use three Signal

Source blocks instead of one, each with a different frequency. The easiest way to do this is to break all the outgoing connections from your current Signal Source. Then copy it and paste it twice. Set the frequency of the first Signal Source to 523.25, the second to 1318.5, and the third to 1568. Next, delete the QT GUI Entry block since we don't need it anymore. When you're done, you'll have something like Figure 5-12.

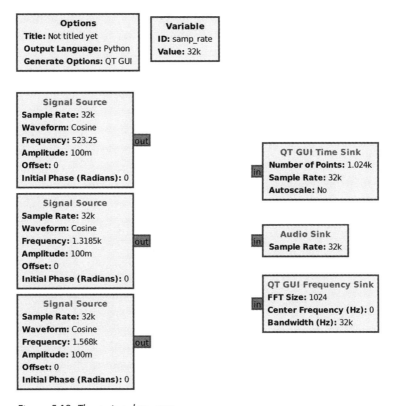

Figure 5-12: Three signal sources

To combine these three tones into one signal, we need to add them together. Find the Add block and add it to the flowgraph (be careful not to use the Add Const block by mistake). It has only two inputs by default. To add a third, change the block's Num Inputs property to 3. While you're at it, change the IO Type to float, as shown in Figure 5-13.

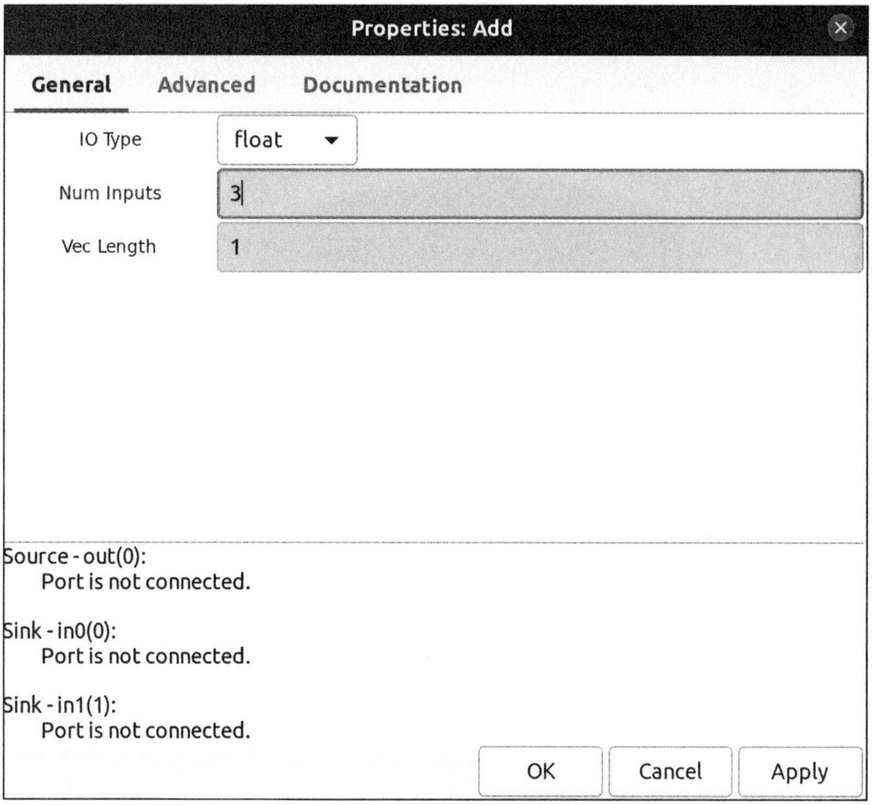

Figure 5-13: The Add block properties

Notice how blocks like Add that have both float and complex functionality (for the Type parameter) all seem to default to complex? This is a clue to the fact that most of the time we'll be working with complex numbers. Slicing deep into the onion to get to the theory of complex numbers will only bring tears at this point (the onion analogy never gets old), so we'll keep to our gradual peeling. But be aware that complex numbers are common in GNU Radio and that we'll discuss them eventually.

Click **OK** to return to the workspace view, and you'll see that the block has sprouted another input. Now connect each of the Signal Source outputs to one of the Add block inputs. Then connect the Add block output to all three of the sinks: the Audio Sink, the QT GUI Time Sink, and the QT GUI Frequency Sink. When you're done, things will look like Figure 5-14.

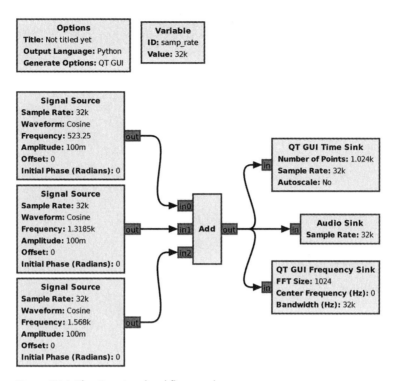

Figure 5-14: The C major chord flowgraph

Execute the flowgraph and you'll hear a richer sound than the previous tone. This isn't shocking, since you're now putting three tones together. For your information, what you're hearing is a C major chord. We've just separated some of the notes by a few octaves to make them easier to distinguish. Figure 5-15 shows what the chord looks like in the time and frequency domains.

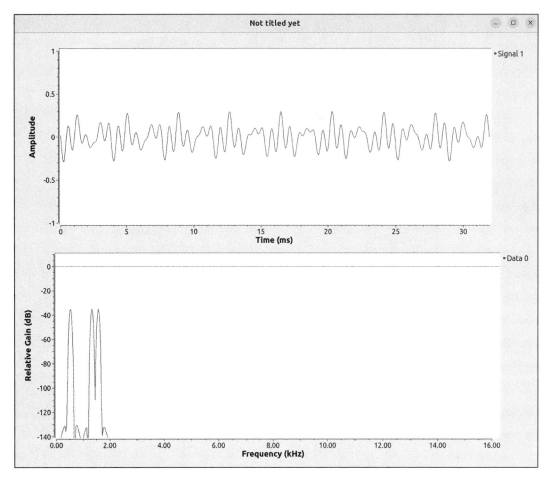

Figure 5-15: The C major flowgraph execution window

On the time domain plot, notice that the waveform is more complicated than the simple sinusoid you saw earlier for the single tone. This con-
firms the assertion that sounds containing multiple frequencies have more
complicated waveforms. On the frequency domain plot, notice that there
are three peaks rather than just one. If you zoom in on the three peaks,
you'll see that they occur around the three frequencies you'd expect, the
ones used for the Signal Source blocks: 523.25 Hz, 1,318.5 Hz, and 1,568 Hz.
You can also see this in Figure 5-16.

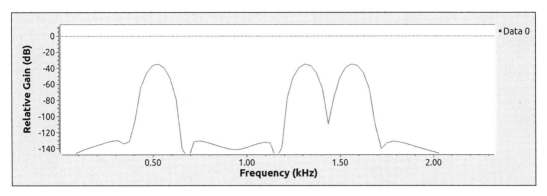

Figure 5-16: A zoomed-in frequency plot of the C major chord

If you'd like to verify the frequencies of each of these peaks (and you should), hover your mouse over any part of the QT GUI plot and you'll see the x and y values at your cursor. In the case of the frequency plot, the x is the value of the horizontal (frequency) axis and will give you a good approximation of the frequency of each peak. The y value displayed over the cursor is something with a unit of dB, short for *decibels*. We'll discuss what decibels mean later in the chapter.

Seeing the three peaks in the frequency plot may not seem like a big deal, since the three tones in the C major chord started out as separate Signal Source blocks, but it's important to recognize what's happened here. When the three tones passed into the Add block, they were merged into a single signal (as evidenced by the resulting complicated waveform in the time domain plot). In the process, the individual frequencies of the three tones effectively became indistinguishable without the help of some mathematics. Specifically, in order to create the frequency plot of this merged signal, the QT GUI Frequency Sink had to conduct an FFT, allowing you to identify the individual frequencies contained in the signal.

Plotting Real-World Sounds

Real-world sounds, like people talking, dogs barking, car engines revving, or even musical instruments playing, are made up of far more than the three frequencies of our computer-generated C major chord. To prove it, we'll analyze a recording of a human voice. Resave your current flowgraph as *voice_fft.grc*. We'll come back to the C major chord project later in the chapter, so don't forget where you put that file.

Next, left-click and drag to select all the Signal Source blocks and the Add block, and press DELETE to get rid of them all. In their place, add a Wav File Source block and connect it to each of the three sinks. This block will inject the audio from an existing WAV file into the flowgraph. Open the block's properties and navigate to the location of the project files you downloaded earlier from *https://nostarch.com/practical-sdr*. Select the one named *ch_05/HumanEvents_s32k.wav*. When you're done, your workspace will look like Figure 5-17.

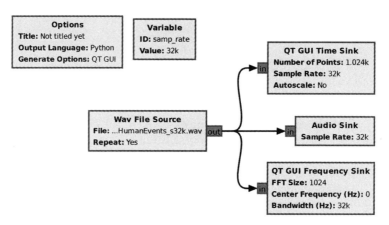

Figure 5-17: A flowgraph to display a voice signal

When you execute this new flowgraph, you'll hear a voice repeating over and over. You'll also see that the time and frequency domain displays are quite a bit busier than those for the simple tone and C major chord. Although the plots bounce around a lot, you can see a representative snapshot of these displays in Figure 5-18.

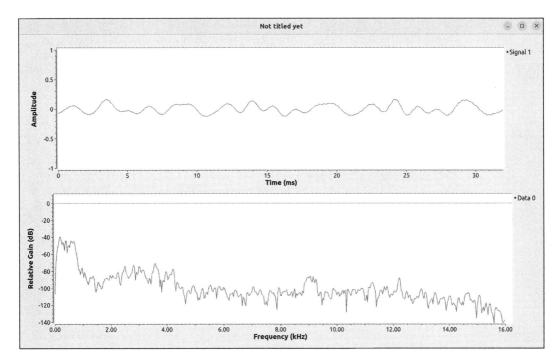

Figure 5-18: The time and frequency plots of the voice signal

The time domain plot shows a great deal of what looks like random movement, an indication that real-world sounds have much more complex waveforms than simple sinusoids, but the more important thing here is the frequency plot. Look at how many peaks the plot has, and notice how spread out the frequencies are in just this simple recording of a human voice. You can see that most of the signal is composed of low to medium frequencies, but there are a few higher ones in there as well. The frequency plot also moves around a fair amount, shifting while jumping up and down.

NOTE *Try recording some sounds of your own in WAV format and see what they look like. You'll need to configure your audio recording software to use a 32 kHz sampling rate so your WAV file will work with this flowgraph.*

It may not be obvious how exactly the FFT gets its job done, but hopefully it's becoming a bit clearer what that job is and how important it is. From now on, when we think about signals, we will often be thinking about the frequencies present in those signals. The more you work with the frequency domain, the more you'll build the intuitions necessary to design your GNU Radio flowgraphs and to debug them when they don't quite do what you intend.

In this section, you've created some flowgraphs allowing you to visually examine some signals. But ultimately you'll want to do more than just look at signals, right? You'll also want to be able to change the signals you encounter with your SDR. There are many ways to process signals, but we'll start by looking at two of the most foundational methods: gain and filtering.

Gain

Applying *gain* to a signal simply means making it bigger, but without changing the signal's shape. For example, if you have a signal with a maximum level of 3, we would say that its *amplitude* is 3 (this will work with any type of unit). If you increase the size of the signal at every point in time so that it now has an amplitude of 6, then your gain is 2. If you were to increase the signal's size from 3 to 9, your gain would be 3. Mathematically, we can express this as follows:

$$\text{gain} = \frac{\text{output size}}{\text{input size}}$$

Don't worry too much about how to compute a specific number for the input size; that can get somewhat involved for complicated waveforms, and we really don't need that kind of detail to build software-defined radio systems. Just consider that applying a gain to a signal will increase that signal at every point in time by that same multiplier.

When it comes to audio signals, applying a gain corresponds to making the audio louder or softer. For radio signals, applying gain will sometimes have a similar effect, but rather than the result being something your ears can sense, the gain will make the signal more prominent to a radio receiver.

Applying a Gain to a Signal

To see firsthand how gain works, we'll use GNU Radio Companion to apply a gain to a signal. Open a new project and save it as *gain.grc*. Next, add a Signal Source, setting its Output Type to Float and its amplitude to 0.01. After that, add a Multiply Const block, setting the IO Type to Float and the Constant property to gain_val. Then add an Audio Sink. Finally, insert two copies of a QT GUI Time Sink, and for both set their Type properties to Float, the Y Min values to -0.1, and the Y Max values to 0.1.

Now that all your blocks have been placed, you can connect them. Start by routing the Signal Source output to the Multiply Const input and the Multiply Const output to the Audio Sink input. Then attach one of the QT GUI Time Sink blocks to the output of the Signal Source and the other to the output of the Multiply Const block. When you're done, your project should look like Figure 5-19.

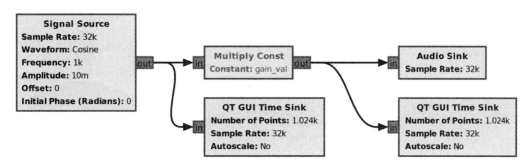

Figure 5-19: The partially completed simple gain project

What's going on here? The Signal Source block generates a small 1,000 Hz sinusoid. It goes into the Multiply Const block, where every point in the signal will be multiplied by whatever the value of gain_val happens to be. This is where you apply the gain to the signal, and thanks to the pair of time sinks, you'll be able to look at the signal both before and after the multiplication process (as well as hear it on your computer speakers or headphones thanks to the Audio Sink). Hardware and software entities whose purpose is to provide gain to a signal are typically referred to as *amplifiers*. Be aware, though, that many other types of nonamplifier blocks may have some gain associated with them, alongside other forms of signal processing.

To make the flowgraph work, we need to provide a value for gain_val, preferably one that can be changed during execution. We'll once again use a QT GUI Range to control the gain with a slider. Set its ID to gain_val, its Default Value to 1, its Start value to 0.1, its Stop value to 10, and its Step value to 0.1. You can leave the other properties alone.

Before running the flowgraph, change the Name property of the first QT GUI Time Sink (the one connected to the Signal Source) to Input, and set the Name of the other QT GUI Time Sink to Output. This will print an informative title on the top of each waveform plot in the execution window. It's generally a good idea to label your instrumentation blocks if you're going to have more than one of them. If all your QT GUI blocks were unnamed, you might find it difficult to determine which signal is contained in which display.

One more thing before running the flowgraph: we'll be making the output signal quite a bit bigger at times throughout this experiment, so it will be easier to adjust the y-axis scale of the instrumentation blocks ahead of time. This way, you won't have to worry about zooming in or out during execution. To do this, double-click each of the QT GUI Time Sink blocks and change the Y Min property to -0.1 and the Y Max property to 0.1. When you're done, you should have something like Figure 5-20.

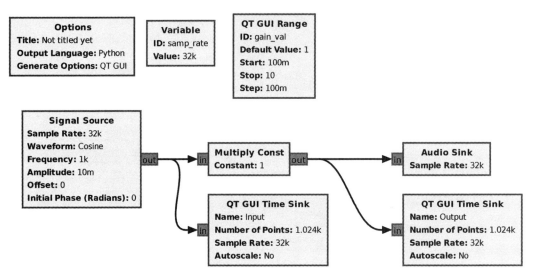

Figure 5-20: The completed simple gain project

After executing the flowgraph, you should hear a tone and also see two tiny but identical sinusoids, like in Figure 5-21.

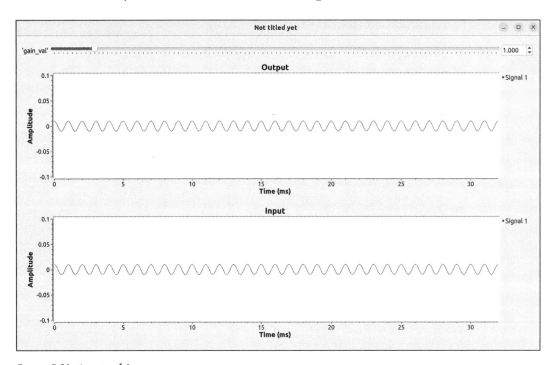

Figure 5-21: A gain of 1

The input and output signals are identical because gain_val defaults to 1, and applying a gain of 1 to a signal doesn't change it. After all, the signal is just a series of numbers, and multiplying a number by 1 doesn't change that number.

Now try changing your gain. You can do this in four ways:

- Click the slider element and drag it left or right.
- Enter a new number into the text box at the right side of the QT GUI Range element.
- Click the up or down arrows next to that text box.
- Click the text box or the slider and then press the up or down arrow keys.

Use one of these methods to increase gain_val to 5. As a result, you should see the output sinusoid grow by a factor of 5, as in Figure 5-22. The tone should also get louder.

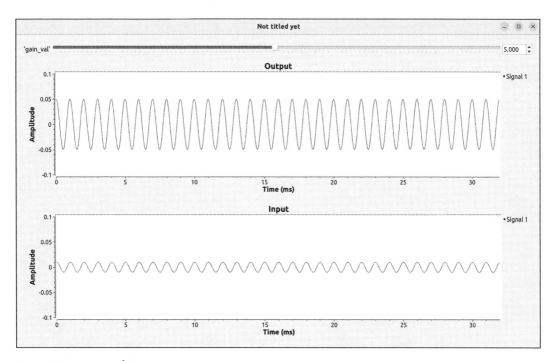

Figure 5-22: A gain of 5

Can you also see how the output waveform shrinks when you move the slider to the left? In fact, you can even make the output smaller than the input signal by setting gain_val to something less than 1. For example, try moving the slider down to 0.3 to see the signal shrink to 30 percent of its original size, as in Figure 5-23. Applying a gain that reduces the size of a signal is typically called *attenuation*.

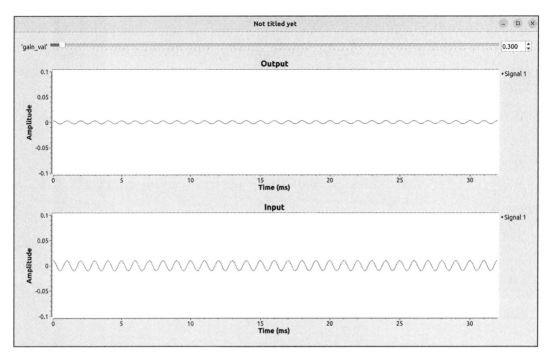

Figure 5-23: An attenuation of 30 percent

If you're feeling adventurous, go ahead and substitute the *ch_05/HumanEvents_s32k.wav* file from the previous section (or one you recorded yourself) for the signal source in the current flowgraph and play around with the gain settings. Although the file doesn't have a steadily repeating waveform like the simple sinusoid, you'll still find that the output waveform's peaks get higher or lower as you turn the gain up or down. You'll also hear the audio get louder and softer as you change the gain. One warning, though: if you set the gain too high, the audio may start to get choppy and distorted, as there are limits to how large of a signal can be sent to your sound card without issues.

Before we move on, consider that in the last few examples you've seen a few different math-related blocks applied to signals, and each one has worked a bit differently. When we generated the C major chord, we used an Add block, which can have a variable number of inputs; in that case, we needed three. This time, to apply a gain, we used a Multiply Const block, which can have only one input. Can you see why one block permits multiple inputs and the other doesn't? In the first case, we were combining 3 different signals, or streams of numbers, by adding them together at each point in time; we could just as easily do the same for 2 signals, or 5, or 20. In the second case, we were concerned only with multiplying one signal, or data stream, by a single constant number defined as a property in the block. There's no place for additional inputs in that process; it's just the one signal being manipulated. If you look at the library of blocks available, however, you'll see that there's also a Multiply block and an Add Const block, essentially the counterparts of the two blocks we used. The `Multiply` block, which can have several inputs, allows you to combine different signals by multiplying the number streams together at each point in time. Meanwhile, the `Add Const` block can have only a single input and is used to add a single constant number to one signal.

Thinking in Decibels

Gain is most often measured in *decibels*, or *dB* for short. Decibels are a logarithmic measurement, similar to the Richter scale used for measuring earthquakes. With the Richter scale, any increase of 1 on the scale equates to a factor-of-10 increase in earthquake magnitude: an 8.0 earthquake is 10 times stronger than a 7.0 earthquake, a 4.0 earthquake is 10 times stronger than a 3.0 earthquake, a 6.7 earthquake is 10 times stronger than a 5.7 earthquake, and so on. Similarly, because decibels are logarithmic, a small increase expressed in decibels can translate to a very large increase in gain.

To explore how decibels work, we'll add some instrumentation blocks that measure in units of dB to the previous gain project and see how those blocks respond to changes in gain. Specifically, we'll add some `QT GUI Frequency Sinks`, since their y-axes are given in dB. But first, resave the project as *gain_db.grc*, then right-click each of the **QT GUI Time Sink** blocks and select **Disable**. Notice that the blocks turn gray when you do this, along with their connections, as shown in Figure 5-24. This feature allows you to make quick changes to a flowgraph that you can easily undo by re-enabling the blocks.

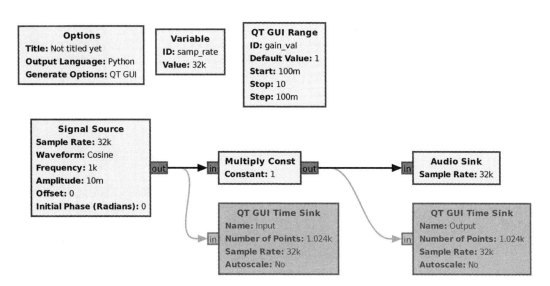

Figure 5-24: The gain flowgraph with disabled `QT GUI Time Sink` blocks

Now add a `QT GUI Frequency Sink` block, changing its Type property to `Float`, its Spectrum Width to `Half`, and its Number of Inputs to `2`. After changing the number of inputs, the block will add a second input port.

Connect the two input ports in place of the disabled time sinks, with in0 connected to the Signal Source output and in1 to the Multiply Const output. While you're at it, change the Stop value of your QT GUI Range to 100. When you're done, you should have something like Figure 5-25.

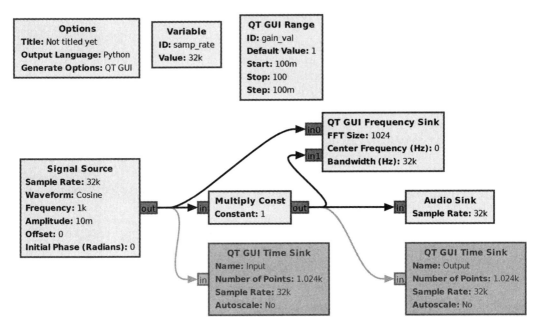

Figure 5-25: The gain flowgraph modified for decibel measurements

You'll now be able to observe the effect of changing the gain in terms of dB. When you run the flowgraph, at first you'll see only one FFT spike because the input and output spikes will be on top of each other. They'll both have a peak at about −55 dB, as in Figure 5-26. (Remember that you can hover your mouse over a peak to get a quick measurement.)

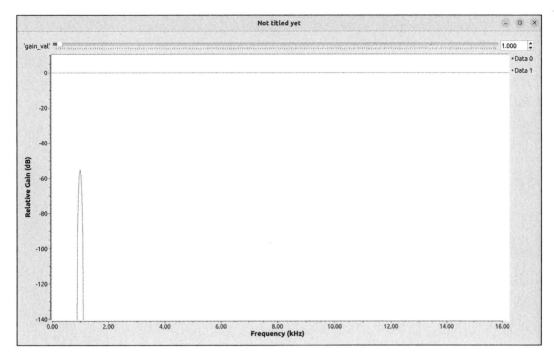

Figure 5-26: A gain of 1 in decibels

We expect the input and output spikes to be the same at this stage; again, multiplying any data by 1 should result in identical data. Now change the gain_val to 10 and examine what happens. The peak labeled Data 1, which corresponds to the output signal, should jump from –55 dB to about –35 dB. This makes for an increase of 20 dB, as shown in Figure 5-27.

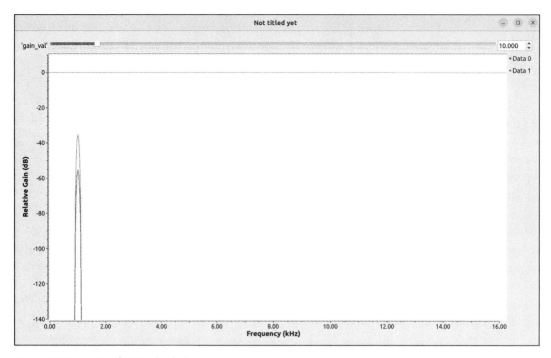

Figure 5-27: A gain of 10 in decibels

What do you think will happen when you change the gain to 100? What about 0.1? Try it!

The takeaway here is that increasing the signal's amplitude by a factor of 10 results in a change of 20 dB on the y-axis of your frequency plot. Put another way, an increase of 20 dB on your plot means you've applied a factor of 10 gain to the original signal. Furthermore, subtracting 20 dB means attenuating the signal to 10 percent of its input value, as you saw when you plugged in a gain value of 0.1 (see Figure 5-28). In that case, the decibel level went from –55 at the input to –75 at the output.

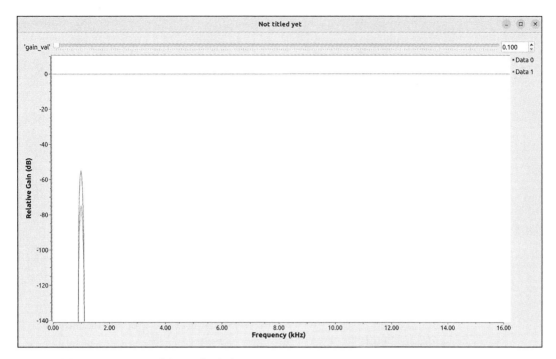

Figure 5-28: An attenuation of 0.1 in decibels

Decibels are a convenient way of keeping track of the gain in a system, especially when several different gain operations are applied to a signal. Consider the diagram in Figure 5-29.

Figure 5-29: A chain of linear gains

Here, a signal passes through four different gains between the input and output: first a gain of 4, then a gain of 500, then an attenuation of 0.025, and finally an attenuation of 0.16. To determine the net result, we have to multiply these linear gain values together. Pulling out your calculator, you'll see that this mess resolves to a simple gain of 8 (where $4 \times 500 \times 0.025 \times 0.16 = 8$).

Now consider Figure 5-30, where the same series of gain operations is rendered in decibels.

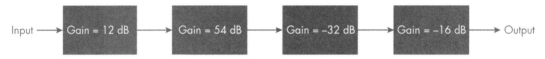

Figure 5-30: A chain of decibel gains

When the gains are expressed in decibels, all we have to do is add them together to determine the net effect, which is much easier than multiplication. In this case, adding the decibel gains of the entire signal chain yields a total gain of 18 dB (where $12 + 54 - 32 - 16 = 18$).

Since gain will be given to you in dB most of the time in the radio world, it's helpful to start thinking in these terms. Down the road, this will be important when you configure the hardware gain of your software-defined radio, as well as if you decide to add some hardware amplifiers to your SDR systems. Gain and decibels also play a central role in filtering signals, which we'll look at next.

DECIBELS IN REAL-WORLD RADIO SYSTEMS

We've established that a factor of 10 (10x) gain corresponds to an increase of 20 dB. However, when working with real-world radio systems, both software-defined and conventional, a gain of 10x is equivalent to 10 dB, not 20 dB. The key to this discrepancy is the distinction between the amplitude and power of a signal.

Radio engineers typically work with power and have defined decibels accordingly: a 10x increase in power corresponds to a +10 dB gain. Signal amplitude, however, isn't the same thing as signal power, and a 10x change in amplitude doesn't correspond to a 10x change in power. Without getting into the physics, just be aware that a 10x change in signal amplitude produces a 20 dB change in power. (If you want to investigate the math and physics, the squaring of voltage to get power is where to start looking. But that's beyond the scope of this book.)

The key takeaway is this: when working in GNU Radio, you'll most often see the 20 dB figure associated with a 10x gain, but later on, when you're working with radio hardware, expect to see the 10 dB power gain figure.

Filters

A *filter* is a processing technique that selectively removes some parts of a signal but not others. Think about a water filter for a moment. You pour impure water through it, and the filter (hopefully) stops most of the impurities from going through, while allowing the water to pass. On the far side of the filter, you have your desired object: pure water.

Real-world radio signals are a lot like that impure water. Ideally, you could put an antenna up in the air and the only signal that it would pick up would be the one you want, but unfortunately, nearly the opposite is the case. You'll get not only the signal you want but also a massive amount of other signals, along with noise and interference (topics we'll explore in Chapter 8). All of this comes into the receiver simultaneously. Filters help isolate just the signal you want, while getting rid of nearly everything else. You put a big mass of radio energy into a filter and design the filter so that it separates what you want (your signal) from the stuff you don't want (other signals, noise, interference, and so on).

Most of the filters you'll use are based on frequency. Earlier we talked about how signals can be broken down into their component frequencies via the Fourier transform. Taking this viewpoint, any given signal will be made up of a number of frequencies, some of which you might want and others which you may not. You can design a filter to reduce the frequencies you don't want, while preserving the ones you do. Another way to think about this is that the filter applies a gain to the signal, but the gain varies depending on frequency. For the frequencies you want to pass through unaffected, called the *passband*, the gain is 1. For the frequencies you want to get rid of, called the *stopband*, the gain is 0—in theory, anyway. In practice, filters aren't quite that perfect, as we'll discuss later in the chapter.

There are four main types of frequency-based filters: low-pass filters, high-pass filters, band-pass filters, and band-reject filters. They vary based on which frequencies they eliminate and which they preserve. We'll consider each kind in turn.

Low-Pass Filters

A *low-pass filter* allows the lower-frequency parts of an input signal to pass through mostly unchanged, but it stops most of the higher-frequency parts of the signal from getting through. Consider the FFT in Figure 5-31, with parts of the signal spread across both low and high frequencies.

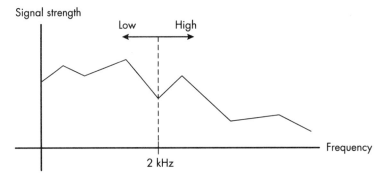

Figure 5-31: An FFT of a signal split into "low" and "high" frequencies

Let's say that any frequencies in this signal above 2 kHz are "high" and any frequencies below that are "low," as Figure 5-31 indicates. Note that there's nothing in physics or mathematics that stipulates where to draw this line between low and high; it's an arbitrary decision based wholly on what we, the filter designers, intend to do. In one context 2.5 kHz may be too high, but in another context it might be exactly the frequency we want. The dividing line between the low frequencies we want and the high ones we don't want is called the *cutoff frequency*.

The goal of the low-pass filter is simply to reduce the part of the signal above the cutoff frequency (2 kHz) as much as possible while affecting the part below the cutoff as little as possible. In a perfect world, the filter would produce the output FFT shown in Figure 5-32.

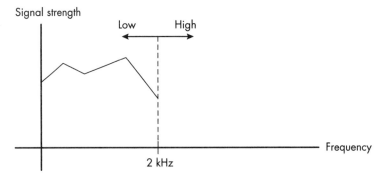

Figure 5-32: An FFT after perfect low-pass filtering

In this ideal low-pass filter, all the frequencies below the cutoff are completely unaffected; they're exactly as strong as they were before. Meanwhile, all the frequencies above the cutoff have been completely removed.

Think about this from an audio standpoint. You're listening to some music but are annoyed by a high-pitched humming sound. When you run your music signal into an FFT, as normal people do when trouble-shooting their audio quality issues, you might see something like the plot in Figure 5-33.

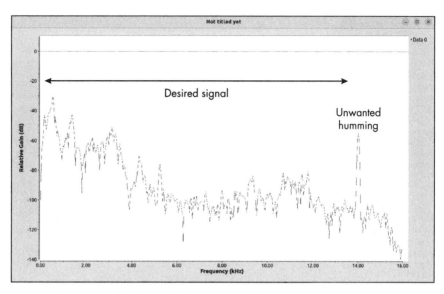

Figure 5-33: An FFT of audio with a humming problem

Notice the large peak in the frequency plot at around 14 kHz. That's the annoying hum, whereas the music mostly ranges from 150 Hz to 12 kHz. Running the signal through a properly designed low-pass filter should be able to eliminate the higher frequency of the hum while leaving the lower frequencies of the music largely unaffected.

To see how a low-pass filter works in practice, we'll return to our C major chord flowgraph from a few sections back (*cmajor.grc*) and try filtering out all but the lowest note. Open the file and resave it as *cmajor_lpf.grc*.

Recall that the three tones in the C major chord have frequencies of 523.25 Hz, 1,318.5 Hz, and 1,568 Hz. Our goal is to add a filter that leaves just the 523.25 Hz tone, while eliminating the upper two tones. Start by breaking the connections between the Add block output and the inputs to the three sinks. Then add a Low Pass Filter block and set its FIR Type to Float->Float (Decimating), its Cutoff Freq to cutoff, and its Transition Width to transition_width (we'll explore these last two properties in detail soon). Then connect the filter between the Add block and each of the three sink blocks. Your flowgraph now consists of an interesting input signal, a filter, and ways to both hear and see the effects of the filter. It should look something like Figure 5-34.

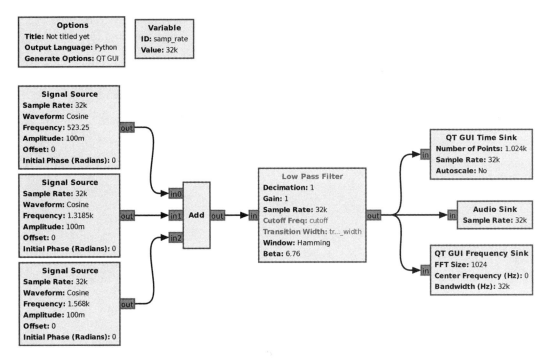

Figure 5-34: The incomplete low-pass filter flowgraph

At this point the Low Pass Filter block should have some red text in it, proof that the flowgraph isn't done yet. We need to assign values to the variables used in the filter. To do so, add a QT GUI Entry with an ID of cutoff, a Type of Float, and a Default Value of 10e3. Then add a second QT GUI Entry with an ID of transition_width, a Type of Float, and a Default Value of 1000. When you're done, the red text will have disappeared and your flowgraph should look like Figure 5-35.

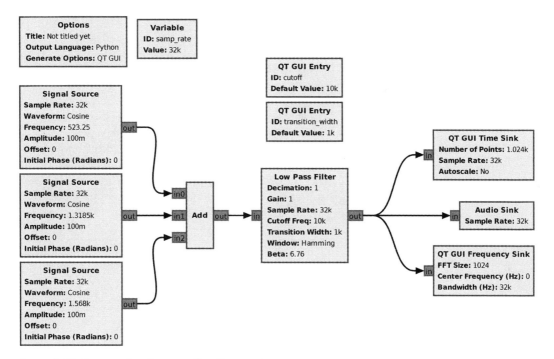

Figure 5-35: The complete low-pass filter flowgraph

When you execute the flowgraph, your ears should tell you that nothing has changed. The frequency plot should also show that all three tones are still there, as shown in Figure 5-36.

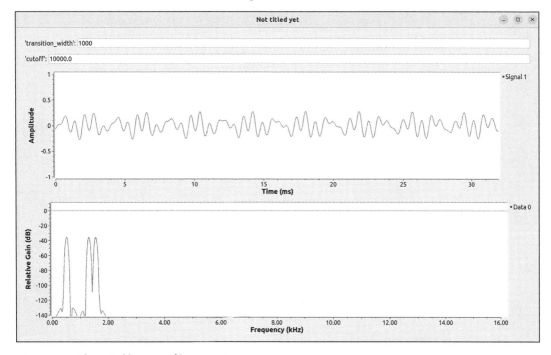

Figure 5-36: The initial low-pass filter output

The signal is unchanged because we set the initial value of the filter's cutoff frequency to 10,000 Hz. Remember, the cutoff is the frequency where the filtering takes effect. In this case, the cutoff is higher than all three frequencies present in the signal, so all three tones have been allowed to pass through the filter.

To isolate the lowest-frequency tone, 523.25 Hz, you'll need to set a cutoff that is higher than that but lower than the next highest frequency, 1,318.5 Hz. (Although we render these numbers with commas, you should not include commas when entering values into the GNU Radio Companion interface or execution windows.) Accordingly, change the cutoff value in the flowgraph execution window to 600 and press ENTER. When you do, the sound should change quite a bit, and so should the two plots of the output signal, as shown in Figure 5-37.

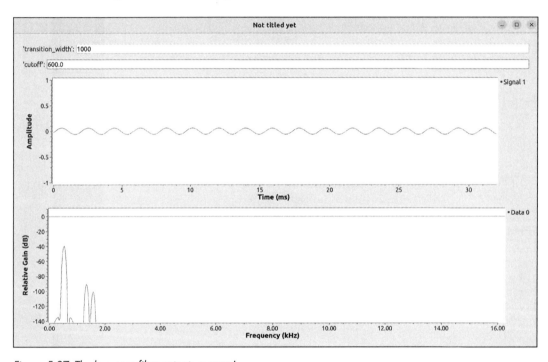

Figure 5-37: The low-pass filter output: success!

You should now hear a single low tone. (Once again, you may need head-phones to make it out.) Looking at the graphical output, the FFT shows the low-frequency peak is intact, while the other two are substantially reduced.

Notice that the higher frequencies haven't been eliminated entirely. However, let me remind you that the y-axis of the FFT display is measured in decibels. The upper two peaks are about 40 dB lower than the first peak. Thinking back to our discussion of decibels, each 20 dB represents a gain factor of 10, so our filter has reduced the unwanted frequencies by about 10 × 10, or 100. In other words, we've reduced the higher two notes of the chord to less than 1 percent of their initial size. The lingering traces of these higher frequencies won't be a problem.

Another useful indicator of the success of the filter is the shape of the waveform in the time domain. As you can see in Figure 5-37, it looks like a clean sinusoid rather than the more complicated waveform of the complete chord. This simple, sinusoidal shape is just what you'd expect if you had indeed filtered out everything but a single tone.

Before moving on, try changing the filter settings to pass the lower two tones while eliminating the highest one. If you have trouble getting this to work, look at *ch_05/solutions/cmajor_lpf2.grc* in the book's companion files for the answer.

High-Pass Filters

A *high-pass filter* reduces the frequencies below a certain cutoff while letting frequencies above the cutoff pass through unchanged, making it the opposite of a low-pass filter. To see how it works, we'll try filtering out all but the highest tone in the C major chord.

Resave your low-pass filter flowgraph as *cmajor_hpf.grc*. Then delete the Low Pass Filter and add a High Pass Filter in its place. Set the filter's FIR Type to Float->Float (Decimating), the Cutoff Freq to cutoff, and the Transition Width to transition_width. Make sure you restore the connections that were removed along with the old filter block, resulting in a flowgraph like in Figure 5-38.

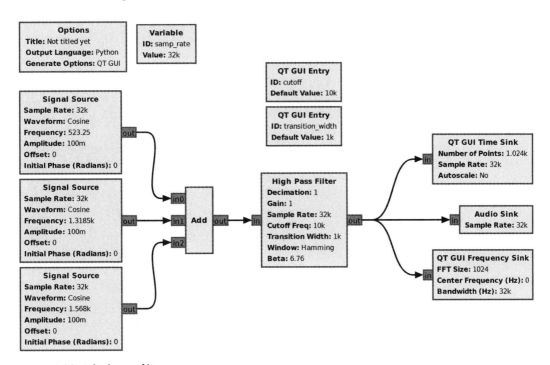

Figure 5-38: A high-pass filter

When you first execute the flowgraph, you won't be able to hear anything, and you should see that all three peaks in the frequency plot have been heavily reduced in size, as shown in Figure 5-39.

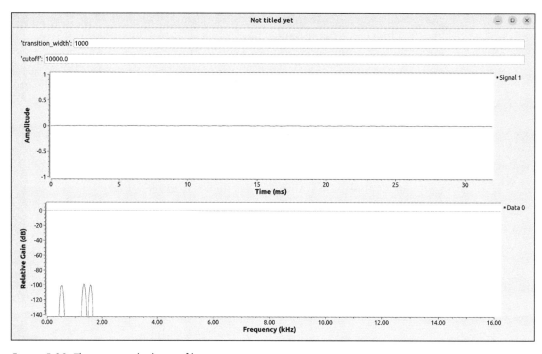

Figure 5-39: The incorrect high-pass filter

All three tones have been filtered out because the filter's cutoff frequency defaulted back to 10,000 Hz. The filter considers anything above that a "high" frequency and allows it to pass through. Conversely, it considers anything below 10,000 Hz (including all three tones in the chord) to be a "low" frequency that should be eliminated.

The tone we want to keep has a frequency of 1,568 Hz, while the next highest tone has a frequency of 1,318.5 Hz. Try changing the cutoff to 1500, just below the highest note, and see what happens. In theory, this should pass the highest note and only the highest note. However, you might notice the suboptimal result shown in Figure 5-40.

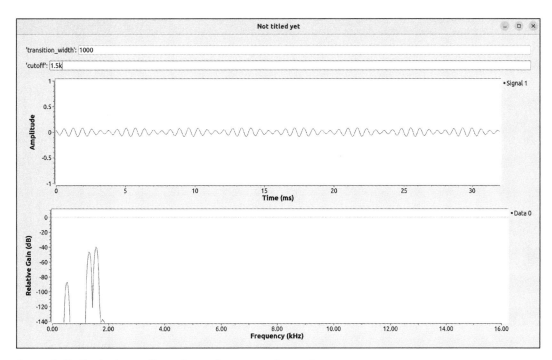

Figure 5-40: The high-pass filter is better, but not good enough yet.

Looking at the frequency plot, although the low note has been reduced quite a bit, the middle note is still pretty strong. If you zoom in, as in Figure 5-41, and hover your mouse over the middle peak, you can see that it's at around −44 dB, only about 7 dB lower than it was on the input side.

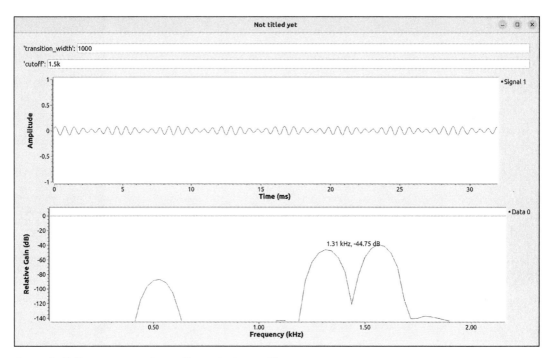

Figure 5-41: Zooming in on the insufficient high-pass filter

You can see more evidence of the suboptimal filtering by looking at the time domain waveform and noticing that the sinusoid is growing and shrinking in size. It isn't perfectly regular, like we'd expect from a pure tone. Depending on your level of musical skill, you may also hear that there's more than one note playing.

You can play around with the cutoff value and improve the filter slightly, but to really eliminate the middle tone, you need to consider another property of the filter: the transition width.

We told you earlier that the frequencies you want to keep are called the passband, while the frequencies you want to eliminate comprise the stopband. I also told you that in practice filters are less than perfect. In fact, there's a third region between the passband and the stopband called the *transition band*, where the filter is partially, but not fully, effective at suppressing the unwanted frequencies. The *transition width* setting of a GNU Radio Companion filter block controls the size of the transition band. The wider the transition width, the less effective the filter will be at reducing frequencies near the cutoff. To illustrate, Figure 5-42 shows a plot of the gain of a filter.

Figure 5-42: The transition width of a filter

In the passband of the filter shown in Figure 5-42, the gain is 1, while in the stopband the gain is roughly 0. In between, in the transition band, the gain gradually changes from 1 to 0. In this transition region, some filtering occurs, but it's suboptimal, and it worsens the closer you get to the cutoff frequency. The farther away from the cutoff, the better the filtering.

Returning to our flowgraph, the default transition width, set by one of our QT GUI Entry blocks, is 1,000 Hz. This means that when we set the cutoff frequency to 1,500 Hz, the transition band ranges from 1,500 Hz down to 500 Hz. The lowest note of the chord (523.25 Hz) is just barely inside the transition band and is filtered very effectively. Unfortunately, the middle note (1,318.5 Hz) is near the high end of the transition band, as shown in Figure 5-43.

Figure 5-43: A signal in the transition band

Since the middle note in the chord falls within the transition band and is much closer to the passband than the stopband, its filtering is especially poor. The best way to fix this is to leave the cutoff frequency at 1,500 Hz and reduce the size of the transition band by shrinking the transition_width value. Any number that creates a transition band that excludes 1,318.5 Hz will work. For example, a transition width of 100 Hz produces the output in Figure 5-44.

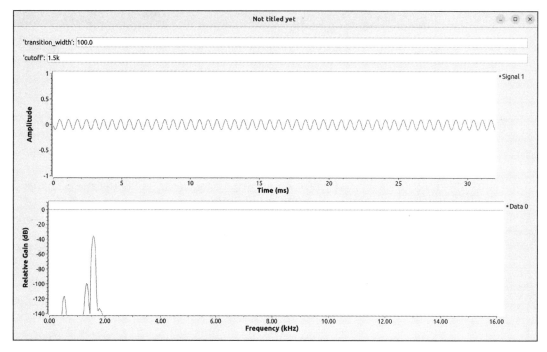

Figure 5-44: The high-pass filter: success!

Now the middle peak in the frequency plot has been lowered about 60 dB, or by a factor of 1,000. Much better! You should also hear a clearer, simpler tone and see a clean sinusoid in the time domain plot.

You might be tempted to choose an extremely small value for the transition width (say, 0.000001 Hz) to create an almost-perfect filter. However, due to the nature of digital signal processing algorithms, the narrower the transition width, the more work the computer must do to implement the filter. A transition width of essentially 0 would take too much computational power to implement. In general, it's not a good idea to make the transition width much narrower than you need.

Band-Pass Filters

A *band-pass filter* has a pair of cutoff frequencies, with the passband between them. Frequencies less than the lower cutoff frequency and higher than the upper cutoff are eliminated (after accounting for the transition bands, anyway). Let's try using one to filter out the top and bottom notes of the C major chord, while leaving the middle note in place.

Save your current flowgraph as *cmajor_bpf.grc*, then replace the High Pass Filter with a Band Pass Filter. When you try to configure its properties, you'll notice something: there are *two* cutoff frequencies. Set them to low _cutoff and high_cutoff. As before, also change the FIR Type to Float -> Float (Decimating), and use transition_ width for the Transition Width property. You'll also need to change the first QT GUI Entry block's ID property from cutoff to low_cutoff and its Default Value to 200 (in Hz). Then make a copy of that QT GUI Entry and change the new block's ID to high_cutoff and its Default Value to 2000. Finally, change the Default Value to 100 for the QT GUI Entry associated with the transition width. If you don't reduce the transition width, you'll get some illegal initial values for your Band Pass Filter block and it won't work correctly, even after you change values in the execution window. When you're finished, your flowgraph should look like Figure 5-45.

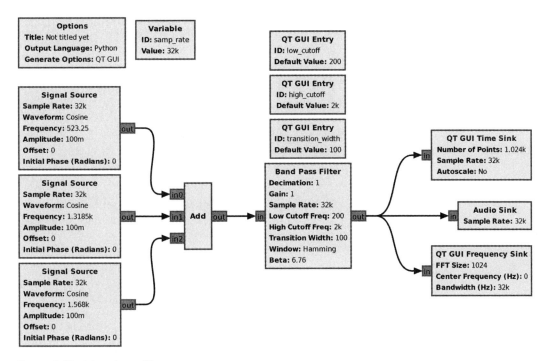

Figure 5-45: A band-pass filter

When you execute the flowgraph, you'll hear all three tones and see all three signals in your FFT plot. If you think about the initial cutoff values you set, this may not be surprising. The passband is between 200 and 2,000 Hz, and all three tones lie within that range. Take a moment and think about how to get your flowgraph working, filtering out the low and high tones while leaving the middle tone intact.

There's more than one way to solve this, but our solution consists of a low_cutoff of 1,000 Hz, a high_cutoff of 1,350 Hz, and the transition_width left at 100 Hz. These numbers produce the output seen in Figure 5-46.

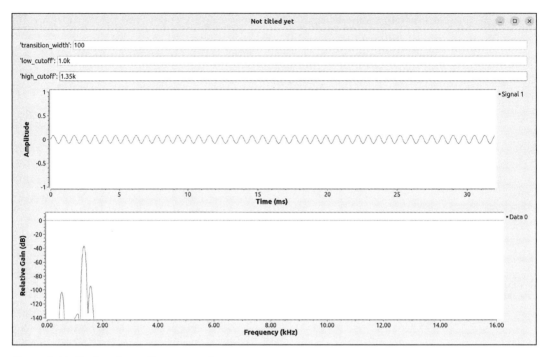

Figure 5-46: The band-pass filter: success!

Notice that the middle peak in the frequency plot has remained high, while the other two are substantially reduced. In the time domain plot, we once again have the simple sinusoidal waveform of a pure tone.

Band-Reject Filters

A *band-reject filter* has two cutoff frequencies. It suppresses, or attenuates, all frequencies between the low and high cutoff frequencies, while preserving those frequencies below the low cutoff and above the high cutoff, making it the opposite of a band-pass filter. In the world outside GNU Radio, band-reject filters may be called *notch filters*.

As an exercise, try using a Band Reject Filter block to eliminate the middle note of the C major chord while leaving the top and bottom notes intact. The Band Reject Filter block has all the same properties as a Band Pass Filter. If you set your low_cutoff to 600 Hz, your high_cutoff to 1,500 Hz, and leave the transition_width at 100 Hz, you should have success. You can also see the *ch_05/solutions/cmajor_brf.grc* file for the solution.

Although you've worked only with audio signals in this chapter, perhaps you can see how useful these filters will be for passing radio signals of interest while stopping signals you don't want to receive. One of the most important parts of a radio is its tuner, and filters are integral to the tuning process, as you'll see in the next chapter.

Creating an Equalizer

An *equalizer* is a system that adjusts the gain of a signal independently in different frequency ranges. If you've ever heard someone talk about "turning up the bass" or "turning up the treble" of some audio, they were talking about using an equalizer. A typical home audio system's equalizer will have several sliders that adjust the volume of the audio, with each slider operating only within a certain frequency range. For example, if you have the leftmost slider pushed to maximum and all the rest to minimum, you'll hear only the lowest (bass) frequencies of your music. The equalizer will apply maximum gain to the low frequencies and zero gain to any frequencies higher than that.

How would you go about building an equalizer in GNU Radio Companion? Assume it should have three sliders that control the low range, midrange, and high range of some audio. Further assume that the low range covers from 20 Hz to 400 Hz, that the midrange is from 400 Hz to 2,600 Hz, and that the high range is from 2,600 Hz to 10 kHz. For input to your equalizer, you can either use the *ch_05/HumanEvents_s32k.wav* from earlier in this chapter or record your own audio.

Note that all the filter blocks we've been using in this chapter have a Gain setting. This setting applies a gain to the output of the filter block. Elsewhere in the chapter, we've left the gain at 1, meaning the passband remains unaltered, but if you set the gain to something higher than 1, the passband would be boosted, even as the stopband is filtered out.

With that in mind, try using filter blocks to build an equalizer in GNU Radio Companion. If you get stuck, take a look at Figure 5-47 or at the *ch_05/solutions/equalizer.grc* flowgraph in your example folder.

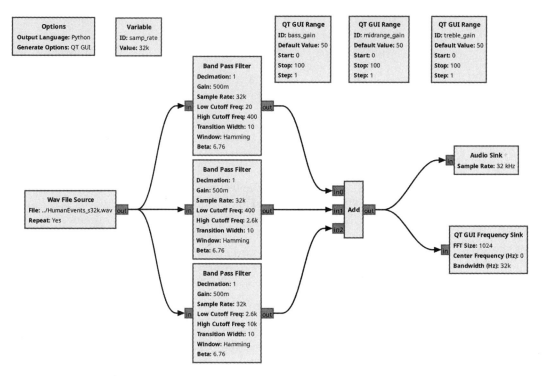

Figure 5-47: An equalizer

The key to creating an equalizer is to join what you've learned about filtering and gain with the technique of combining signals with the Add block. The three filters separate the input into low, mid, and high ranges. Each of these ranges has a unique gain applied to it, allowing you to make it more or less prominent. Finally, the three ranges are combined through addition and sent to the Audio Sink for listening and QT GUI Frequency Sink for viewing.

Conclusion

In this chapter, you learned more about what frequency is and saw how to use FFTs to view signals in the frequency domain. You learned how to apply gain to a signal and how gain can be measured in decibels. Finally, you learned how frequency and gain both play a role in filters, processing techniques that eliminate some frequencies from a signal while preserving others.

Now that you have these concepts under your belt, you have the tools to start dissecting the AM receiver you built in Chapter 4, which is exactly what we'll do next. Although that AM radio may not have seemed like such a big deal, you'll find most of its components in any other radio receiver you eventually build in GNU Radio Companion.

6

HOW AN AM RECEIVER WORKS

In this chapter, we'll take a closer look at the AM radio you built in Chapter 4. With the last chapter's lesson in signal processing under your belt, you're now better equipped to understand how it works. You'll learn how the receiver is able to tune to a particular radio signal in the input data, how that signal is demodulated to extract an audio signal, and how that audio signal is resampled so it can be passed to your sound card.

Much of what we'll be doing in this chapter will involve adding QT GUI sinks to various points in the flowgraph to get a look at what's happening to the radio data at each point. To get started, find the *am_rx.grc* AM radio project we built in Chapter 4, and create a copy named *second_am_rx.grc*. You can do this by opening the file in GNU Radio Companion and saving it with the new name, or you can copy the file in the Linux filesystem. To recap, Figure 6-1 shows how the flowgraph looked at the end of Chapter 4.

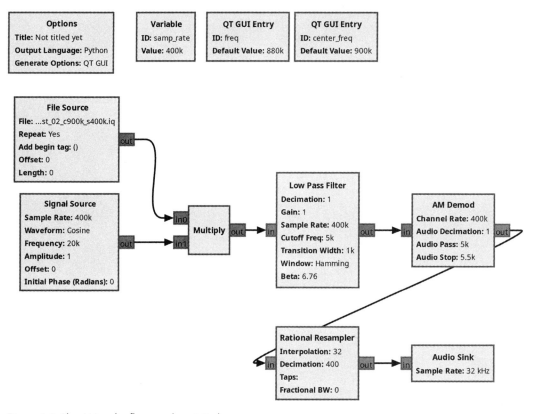

Figure 6-1: The AM radio flowgraph revisited

Over the course of the chapter, we'll start from the beginning, or the left side, of your flowgraph and follow the signals as they flow through to the Audio Sink block on the right. We'll begin by looking at the original source data.

Examining the Input Radio Frequency Data

The input RF data can show you a lot about radio signals if you examine it with the right tools. Add a QT GUI Frequency Sink to your flowgraph and connect it to the output of the File Source. Remember, radio data is entering our flowgraph from this block via a file filled with captured RF data. This new QT GUI sink will allow you to see the frequencies contained in the radio data. Continuing the good habit of helpful labeling, set the new block's Name to RF Input. When you're done, your flowgraph will look like Figure 6-2.

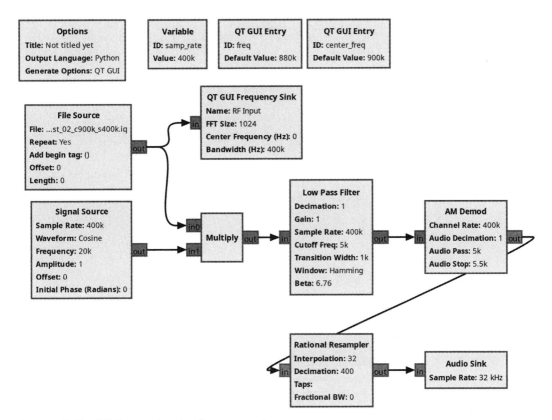

Figure 6-2: The AM flowgraph with a frequency sink

Notice that the output tab on the File Source block is blue, indicating that the data coming from our input file is of complex type. Therefore, unlike in many of the previous exercises, you don't need to change the Type of the Frequency Sink to Float. This isn't simply a special case; in general, all radio data is complex. You'll learn why in Chapter 11, but for now, recognize that your radio data will start out complex and typically transition to a floating-point format as you move closer to your sinks.

Execute the flowgraph to take a look at a fast Fourier transform of the radio data. You'll see something like Figure 6-3.

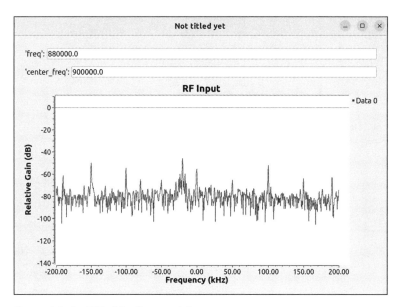

Figure 6-3: An FFT of the flowgraph input

The FFT shows that lots of frequencies are contained in the RF data, but if you look closely, you'll notice that a few peaks stand out above the rest, occurring at suspiciously even frequency intervals. This kind of spacing is what we'd expect to see in raw AM radio data: several stations transmitting at the same time on different frequencies and creating sharp peaks. We'd also expect those peaks to occur at multiples of 10 kHz, just like the AM radio stations that broadcast in your city. After all, each AM station's number, such as 880 or 750, corresponds to that station's broadcast frequency in kHz (880 kHz or 750 kHz).

The evenly spaced peaks make sense, but there's something funny going on with the plot's horizontal axis: it shows that the data contains frequencies ranging from −200 kHz to 200 kHz. Even if we set aside the question of what on earth a negative frequency is, this range doesn't make much sense. You might recall that the default value of freq in the flowgraph is 880e3, meaning you're tuning to a frequency of 880 kHz when you execute the flowgraph. Why aren't we seeing higher frequencies like 880 kHz in the plot?

To understand what's going on, we should first explain something about the radio data we're getting from the File Source block. Its frequency range doesn't extend all the way down to 0 Hz, as you might expect. The frequencies are much more limited than that. The file contains data only for the range from 700 kHz to 1,100 kHz. This range might make a little more sense if we tell you that it's 900 kHz ± 200 kHz and if we remind you that our flowgraph has a variable called center_freq that we set to 900e3.

The original hardware that captured the data can receive it only over a limited range of frequencies, rather than all frequencies starting from 0 Hz. The frequency range for which a radio receiver can grab data is called its *input bandwidth*. In this case, the receiver had an input bandwidth of 400 kHz. When this RF data was originally captured, the input bandwidth was centered at 900 kHz, resulting in captured data from 700 kHz to 1,100 kHz. We've set center_freq to 900e3 to match the flowgraph to the conditions that existed when the data was captured.

Returning to our FFT, notice that the range of frequencies shown, –200 kHz to 200 kHz, correctly corresponds to our 400 kHz input bandwidth. However, the center frequency is given as 0 Hz rather than 900 kHz. This discrepancy has to do with the nature of the sampled radio data. Even though it might seem as if the center frequency of the received radio signals should somehow be embedded in the raw data, it's actually not.

For now, realize that you need to provide a frequency reference to the QT GUI Frequency Sink so that it knows where the center frequency is. You could do that by changing the block's Center Frequency (Hz) property to 900e3 (900 kHz). An even better idea is to change the property to center_freq, so the block will automatically adjust to any changes you ever make to that value. In either case, this change will redraw the horizontal axis of the display without changing the flowgraph's data in any way. When you make the change and rerun, you'll see something like Figure 6-4.

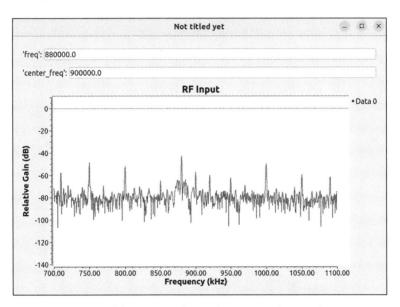

Figure 6-4: An FFT of the RF input data with a corrected center frequency

Now you have an accurate representation of your raw radio data relative to the horizontal axis. We can see frequencies ranging from 700 kHz to 1,100 kHz. Notice how one of the peaks occurs at 880 kHz. This corresponds to the frequency of the station to which you're currently listening, the one specified by the value of freq. Additionally, each of the peaks in the display corresponds to a different broadcast transmission to which you can tune.

Take a close look at this plot, as it's enormously informative and something you'll be using in nearly every radio receiver that you ever design. The plot shows you how many signals are present in the capture range and the frequencies they occupy. The magnitude of this plot at any given point on the x-axis shows you how much RF energy is present at that frequency: larger spikes correspond to more powerful signals. The plot also reveals useful information about noise levels and signal bandwidths, but we won't be ready to talk about those until Chapter 8.

Above all, thanks to our FFT, we're not limited to haphazardly spinning a dial back and forth to find a station. Instead, we can tune directly to any of the peaks we see. In fact, go ahead and tune to some of the other peaks right now by changing the freq value. We'll look at how this tuning process works next.

Tuning

You can think of tuning as focusing in on one specific signal while excluding any others. This is what happens when you operate an AM radio. You spin the dial (or press the digital buttons) until you get the channel for a specific station. The radio then produces the audio for that station, while ignoring all the other channels that might be out there.

Your AM receiver flowgraph implements tuning as a two-step process. First, it shifts the input radio signals so that the one you want is centered at 0 Hz in the frequency domain. It does this by multiplying the input data by a sinusoid of a certain frequency. Next, it filters out anything that's *not* the zero-centered signal you want. To illustrate, imagine the data an SDR receives at any given moment. The FFT might look something like Figure 6-5.

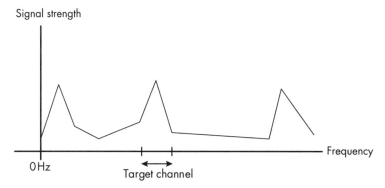

Figure 6-5: An FFT of input radio data

The input data contains three different signals, each represented by a spike in the FFT. Say you want to tune to the middle spike, which we'll call the target signal. You would first shift the frequency of your input radio data so that the target signal is centered around 0 Hz, as shown in Figure 6-6.

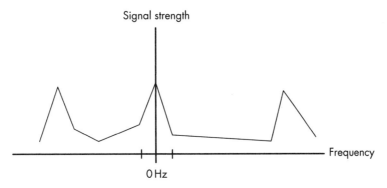

Figure 6-6: Shifted radio data

Notice that the shape of the FFT hasn't changed at all. It's simply shifted position along the x-axis of the plot so that the target signal's peak is located at exactly 0 Hz, rather than at some higher frequency. After making this shift, you would filter everything but the zero-centered target signal, leaving the FFT shown in Figure 6-7. This is your tuned signal! Everything else has essentially been eliminated.

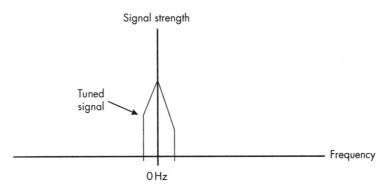

Figure 6-7: Shifted and filtered radio data

Tuning is an essential part of nearly every radio you'll ever build, so it's crucial that you have a firm understanding of how it works. For the rest of this section, we'll conduct some experiments in GNU Radio Companion to give you a hands-on understanding of the tuning process. We'll look at each of the two steps, frequency shifting and filtering, in turn, then put them together to see how tuning happens in your AM receiver flowgraph.

Frequency Shifting

We'll begin with the first step in the tuning process: frequency shifting. As you'll see in this section, the way to shift the frequency of some radio data is to multiply it by a complex-typed sinusoid. We'll demonstrate how this works with a simple experiment in GNU Radio Companion. Set aside your AM radio for now, and create a new project called *freq_shift.grc*.

GNU Radio Companion can have multiple projects open at the same time in different tabs, so there's no need to close your AM receiver before creating this new project.

In your new project, drop down a `File Source` and set its File property to *ch_06/rf_input_c0_s32k.iq*. This file, included in the project files accompanying this book, contains synthetic radio data that you'll shift around in the frequency domain using tuning techniques. Next, add a `Throttle` block and connect its input to the output of the `File Source`. We'll explain more about this block in a moment, but in short, it will keep your computer from working too hard. Then, add a `QT GUI Frequency Sink` and connect its input to the `Throttle` block output. This way you'll be able to look at the data coming out of the input file before you do any processing. Finally, add some appropriate content to your `Options` block, after which your flowgraph will look something like Figure 6-8.

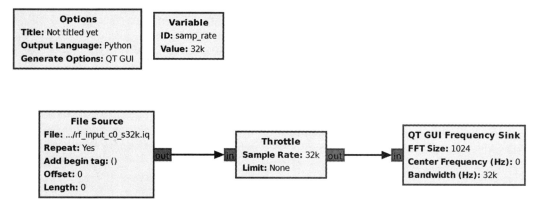

Figure 6-8: A partial flowgraph for the frequency shifter

When you run the flowgraph, you'll see a single peak in the frequency domain, as in Figure 6-9.

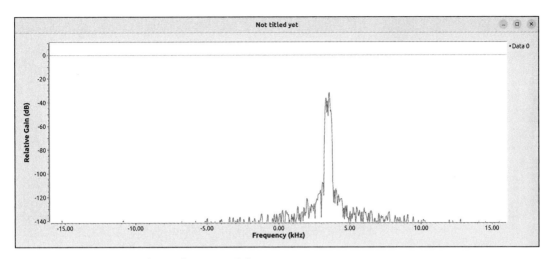

Figure 6-9: Raw input data for the frequency shifter

Next, we'll want to multiply the file data by a sinusoid with a variable frequency. You may recall that we've already created a variable-frequency sinusoid in our AM receiver flowgraph using a `Signal Source` block and a `QT GUI Entry` to control the block's frequency property. We'll do the same thing here. Add a `Signal Source` with a Frequency set to `freq`. Leave all the other properties at their default values. Notice that this means your sinusoid will be of complex type, rather than the floating-point sinusoids you've seen previously. You'll need to define the `freq` value, so do that with a `QT GUI Entry`, setting the ID to `freq` and leaving the Default Value at `0`.

Now that we have a variable-frequency sinusoid, we need to multiply it by the input data. Add a `Multiply` block, connecting one of the inputs to the `Throttle` output and the other to the `Signal Source` output. Next, add a second `QT GUI Frequency Sink`, connect it to the `Multiply` output and give it a Name of `Shifted`. This will allow you to see the results of the multiplication. Finally, change the Name property of the first `QT GUI Frequency Sink` to `Input`. When you're done, the flowgraph should look like Figure 6-10.

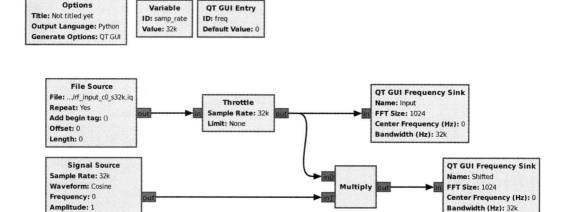

Figure 6-10: A complete frequency shifter flowgraph

Let's take a step back and think about what's happening in this flow-graph. You have some RF data streaming in from a file, each sample of which is being multiplied by a sample of a sinusoid. The sinusoid has an initial frequency of 0 Hz, but that can be changed at runtime. To view the results of all this, you have a pair of frequency plots to show you the RF data both before and after the shifting process. Go ahead and view the results right now by executing the flowgraph. It should produce a plot like Figure 6-11.

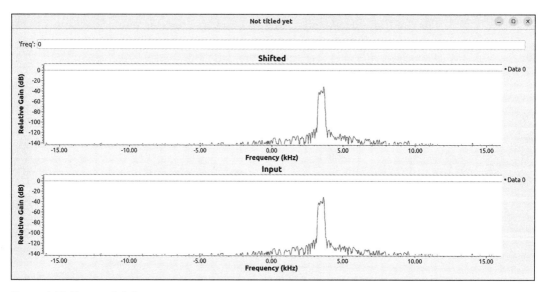

Figure 6-11: The initial shifter output

Not a lot of change, is there? Multiplying by 0 Hz leaves the RF data as is. Now try changing the value of freq to 1000. Figure 6-12 shows the result.

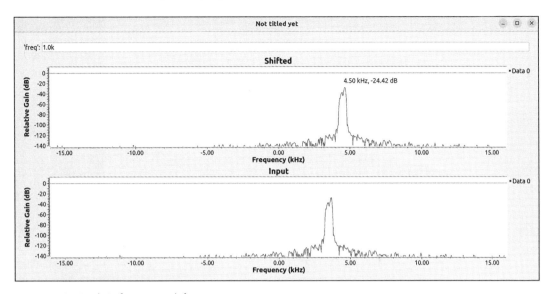

Figure 6-12: A 1 kHz frequency shift

See what happened? If you hover your mouse over the peaks in both plots, you'll see that the input peak is centered around 3.5 kHz, while the shifted peak is around 4.5 kHz. Feel free to try plugging in other frequency values for the sinusoid (they must be integers, since we left the Type property of the QT GUI Entry as Integer). What you'll see is that the frequency plot of the input shifts an amount equal to the frequency of the sinusoid.

This phenomenon isn't specific to the signal we have here; it's actually a general principle. You can take any complex RF data, multiply it by a complex sinusoid of frequency *f*, and you'll get the same complex RF data shifted in frequency by *f*. We won't get into the mathematics of how this works, but the principle holds true regardless of the frequencies involved. Multiply by a 4 kHz complex sinusoid, and the frequency shifts 4 kHz to the right. Multiply by a 9,341 Hz signal, and it shifts 9,341 Hz to the right. Multiply it by a –1 kHz signal, and it shifts 1 kHz to the left.

Wait, did we say –1 kHz? Yes! We know the idea of negative frequencies can be hard to visualize, but we're going to ask you to take their existence on faith as mathematically valid and to accept that multiplying RF data by a sinusoid with a negative frequency produces a leftward shift in all the frequencies present. And that's how we'll shift our peak frequency of 3,500 Hz down to 0 Hz: we'll multiply by a negative frequency. Specifically, plug -3500 into the freq box, and you'll get the output you see shown in Figure 6-13. Notice that the peak of the shifted FFT is now centered around 0 Hz.

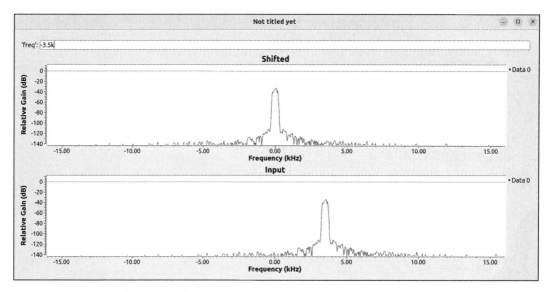

Figure 6-13: A centered frequency plot

To generalize what we've just seen, the first step of tuning to a frequency is to multiply the input RF data by a sinusoid whose frequency is –1 times the frequency to which you want to tune. Want to tune to a 5,400 Hz signal? Use a sinusoid with a frequency of –5,400 Hz. Want to tune to a –4 MHz signal? Use a +4 MHz sinusoid. Again, try not to think too much about what the negative frequencies actually represent right now and just think of them as a useful mathematical tool.

Filtering

The second step in the tuning process is to take your frequency-shifted RF data and filter out everything but your target signal. That brings us to our old friend the filter. As we discussed in Chapter 5, filters eliminate frequencies you don't want, while passing (or preserving) ones that you desire.

Thanks to the frequency-shifting operation we just discussed, you now have the frequencies you want centered at 0 Hz. This should be good, right? You just need to pick a filter that passes the frequencies near zero and eliminates all the other ones. But wait a minute; this isn't quite the same as our previous filter exercises. You now have negative frequencies, some of which you'd like to preserve. What will a filter do with those?

It turns out that filters operating on complex data, like the data you have here, will affect negative frequencies in the same way as positive ones. For example, a low-pass filter will pass frequencies between 0 Hz and the cutoff frequency, and it will also pass frequencies between 0 Hz and –1 times the cutoff frequency. Meanwhile, it will remove frequencies greater than the cutoff frequency and less than –1 times the cutoff frequency. This is exactly what you need. After the frequency shift, you have a peak centered at 0 Hz, with a little bit of your signal on the left side of zero and little bit on the right side. All you need to do is send the data through a low-pass filter and pick the right cutoff frequency.

Start with your frequency-shifting flowgraph and rename it *tuner.grc*. If all goes well, this flowgraph will have both components of the tuner, the frequency shifting and the filtering. Change the File property in the File Source to *ch_06/tuner_test_c0_s1M.iq*. This file will be a bit more interesting than the last one, but it was also captured at a different sample rate (if you're curious, you can see that we embedded it into the filename itself with *s1M*). To account for this new sample rate, change the Variable with ID of samp_rate to have a Value of 1e6. Then add a Low Pass Filter with its input connected to the output of the Multiply block, and a QT GUI Time Sink connected to the output of the Low Pass Filter. We embedded some test patterns in the input file data that will be visible in the time domain, so this new sink will let you see if your tuner is working correctly. The test patterns will be relatively small, however, so to see them clearly, we'll configure the QT GUI Time Sink to automatically set the y-axis of the plot by changing the Autoscale property to Yes.

Next, we have to set the properties of the low-pass filter. Since you don't yet know what kind of value to use for the cutoff frequency, it's a good idea to use a QT GUI Entry so you can adjust the value as the simulation runs. Go ahead and create a QT GUI Entry with an ID of cutoff, a Type of Float, and a Default Value of 100e3. Then double-click **Low Pass Filter** and set the Cutoff Freq to cutoff.

You also need to set the filter's transition width. You could create a separate QT GUI Entry for this, but we're going to let you in on a little secret: when you're doing a lot of laboratory SDR work, where the signals are pretty clear, you don't have to be all that careful with your transition width. As a rule of thumb, you can start off by setting it to one-tenth of your cutoff frequency and change it only if you run into problems. The one-tenth will usually give you decent filtering without overworking your CPU. As such, go ahead and set your Transition Width property to cutoff/10, just as in Figure 6-14.

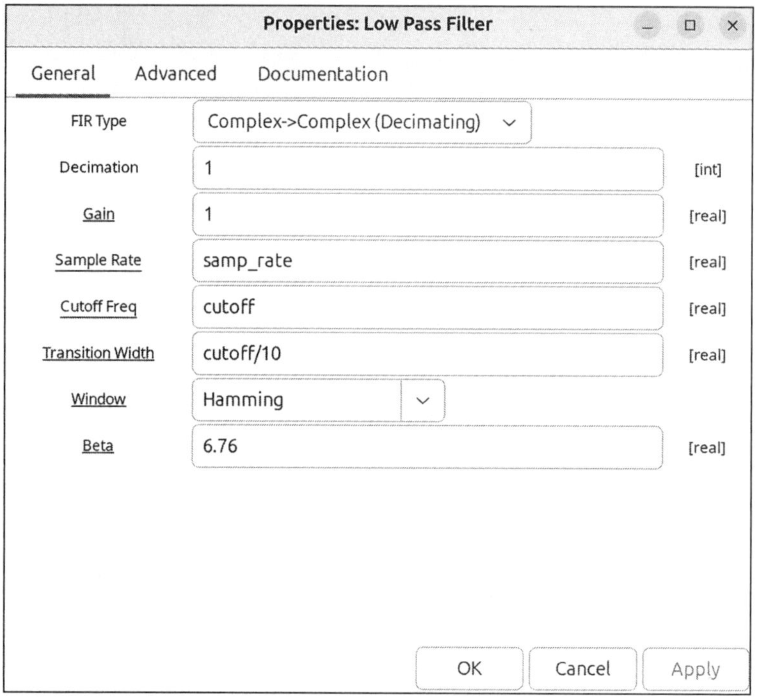

Figure 6-14: The `Low Pass Filter` *properties*

Entering in that simple expression may seem like a small thing, but it means you can change the filter's behavior with a single operation (updating the value of `cutoff`), rather than adjusting both the cutoff and transition width every time you want to try something new. As we'll further discuss in Chapter 7, designing your flowgraphs with these types of expressions makes them much easier to operate.

Finally, let's revisit the frequency shifting portion of the flowgraph for a moment. Right now, the sinusoid has a frequency set directly by the `freq` value, which shifts the RF data to the left (for negative values of `freq`) or the right (for positive values of `freq`). Rather than specify the amount you want to shift the RF data, however, it would be more intuitive if you could just specify the frequency you want to tune to, then have that frequency shift to the center. This is essentially the reverse of what you're currently doing. To make it happen, first change the ID of the `QT GUI Entry` from `freq` to `tune_freq`, then change its Type to `Float`. (You'll be using some bigger numbers for tuning now, and you'll see in a moment that some floating-point notation simplifies entering these larger numbers.) Then change the Frequency property of the `Signal Source` to `-1 * tune_freq`. This way, entering in a value for `tune_freq` will cause the data to recenter on that frequency. When you're done, your flowgraph will look like Figure 6-15.

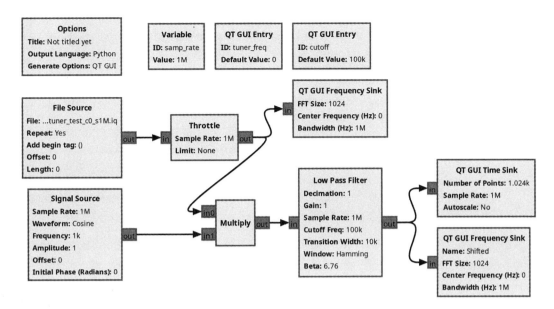

Figure 6-15: A tuning flowgraph

Upon execution, you'll see that the new input file contains three peaks, each with a different width (Figure 6-16). You can also see that the time domain plot just shows waveforms moving around randomly, without the consistent shape we might expect in a test pattern. This makes sense, since we haven't tuned to anything yet.

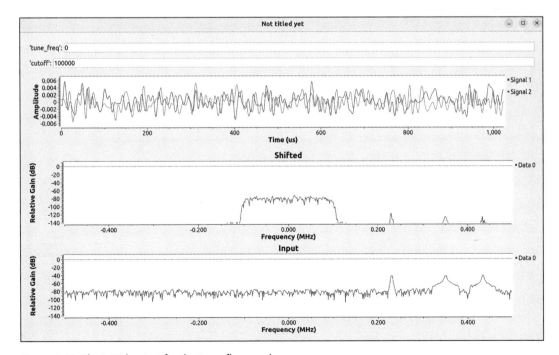

Figure 6-16: The initial output for the tuner flowgraph

Let's try to adjust the tune_freq and cutoff values to tune to the three different signals. The first thing to do is take note of some information about the input RF data. By hovering your mouse over each signal peak, you can see that the three peaks in the frequency plot have frequencies of 230 kHz, 350 kHz, and 435 kHz. You can also use your mouse to get a pretty good estimate of the width of each peak. The first peak seems to be about 20 kHz wide, while the other two are about 60 kHz wide.

We'll tune the lowest frequency first. In the flowgraph execution window, enter **230k** into the tune_freq box. After hitting ENTER, observe that the Shifted waveform now has the first peak centered around zero, as shown in Figure 6-17. If you get an error in the console area of GNU Radio Companion when you press ENTER, make sure you set the Type of the QT GUI Entry for tune_freq to Float.

NOTE *It's a quirk of GNU Radio Companion that block properties use exponential notation, while values entered into QT GUI widgets during execution use metric units like k, M, G, and the like. That's why you entered 230k into the tune_freq box rather than 230e3.*

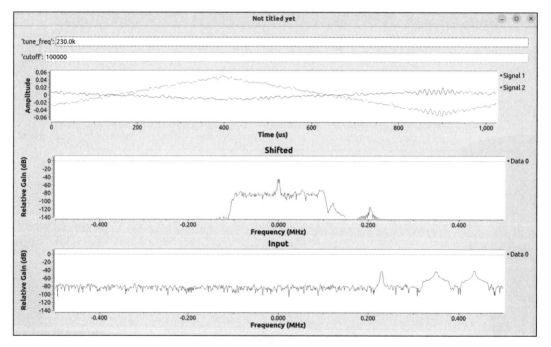

Figure 6-17: Centering the first peak

You should see a triangular-shaped signal appear in the time domain plot, but it doesn't look very clean, so you'll also need to adjust your cutoff frequency. As noted, the first peak seems to be 20 kHz wide, so you might think that 20 kHz is a decent cutoff value. Remember two things, though: the signal peak is centered at zero, and the complex low-pass filter passes frequencies from the −cutoff to the +cutoff frequencies. Therefore, it would be better to use 10 kHz as the cutoff, in which case the filter will pass everything from –10 kHz to +10 kHz, for a total of 20 kHz. With very clear signals, it's not always catastrophic to have a wider filter than necessary, as you'd end up with the same result if you used 20 kHz for the cutoff, but it's important to understand where the numbers are coming from.

Once you set the cutoff to 10k, the triangular wave should clean up significantly, as shown in Figure 6-18.

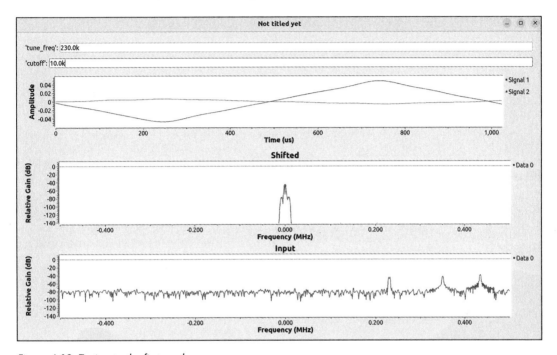

Figure 6-18: Tuning to the first peak

The time domain signal you're looking at is a complex triangle wave, which we're using as a simple test pattern to show you that the tuner is functioning correctly. It's only possible to see one cycle of this waveform, though, and it would be nicer to zoom out on the horizontal axis to see more of it. To do this, click your middle mouse button anywhere in the time domain plot to bring up a context menu with numerous options. Click **Number of Points**, and you'll be shown a dialog box populated with 1024. Changing this to a larger number will display more samples on the plot, essentially zooming out in the horizontal direction. For example, try doubling the number by entering 2048, and you'll get a wider view of your signal in time, as shown in Figure 6-19.

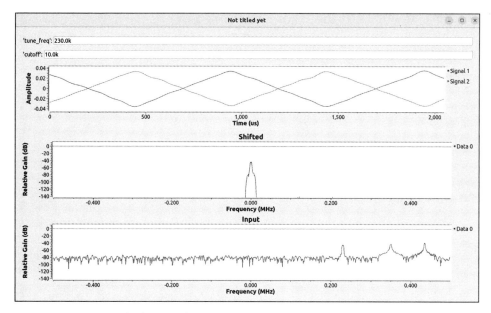

Figure 6-19: Tuning to the first signal

Next, tune to the second signal by changing tune_freq to 350k. The time domain waveform will change to something with a sharper upward ramp and a slower downward ramp than the triangle wave (Figure 6-20).

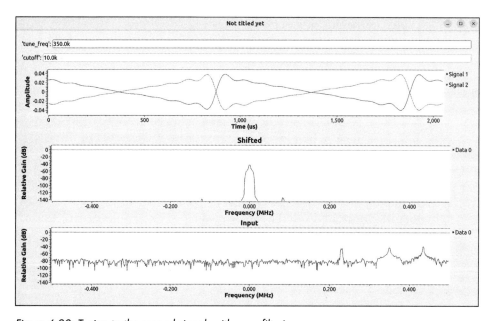

Figure 6-20: Tuning to the second signal, with poor filtering

As you may recall, however, the second and third peaks are wider than the first, so you'll need to update your cutoff value. When you change it to

30k (half of the peaks' 60k width) and press ENTER, you'll see the waveform get a bit sharper, as shown in Figure 6-21.

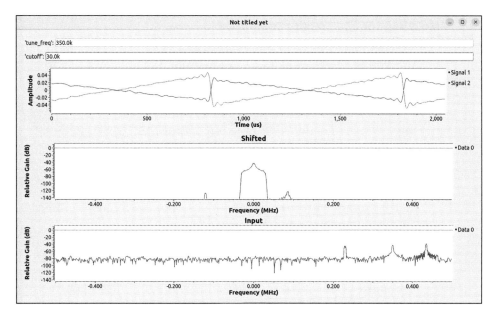

Figure 6-21: Tuning to the second signal, with better filtering

This waveform is known in electronics as a *sawtooth wave*, with a sharp rise and a slow ramp-down.

Finally, change the tune_freq to 435k to see the last signal's test pattern (Figure 6-22).

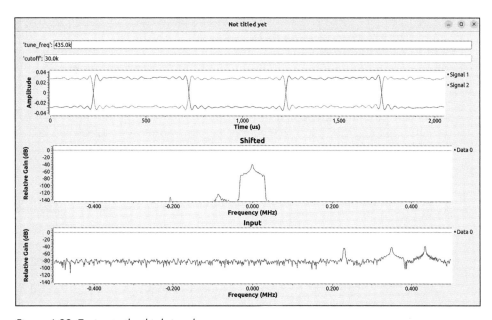

Figure 6-22: Tuning to the third signal

Your last waveform is a complex square wave. Feel free to play around with the tuning and cutoff values to get a feel for what they do.

Accounting for Real-World Frequencies

When you're working with real-world radio data in an SDR flowgraph, as in your AM receiver project, you have to keep two sets of frequencies in your head: those corresponding to real-world physics and those that GNU Radio sees. This adds a complication to the tuning process for real-world data, in contrast to the tuning exercises we've just worked through.

Take a look back at the input FFTs in Figures 6-9 and 6-16. In both of these exercises, the input files had frequency plots ranging from negative frequencies through zero and on to positive frequencies. Despite our reassurances about negative frequencies being mathematically viable, SDRs don't really capture RF data at negative frequencies. The actual radio signals you'll encounter in the real world will have positive frequencies that are typically quite a bit greater than zero. We saw this in the frequency plot of the input radio data for the full AM receiver (Figure 6-4). Once we adjusted the horizontal axis of the plot, it showed us that the data starts at 700 kHz and goes up to 1,100 kHz. Unlike the zero-centered plots, this is representative of what's going on in the physical world. The frequency values shown in Figure 6-4 match the actual physics in play at the time the file was captured. For example, the signal you're tuning to by entering 880e3 (or 880000 as it's displayed in the execution window) was actually present at 880 kHz when we received it with our SDR and streamed it to the file.

Once those RF signals are captured by the SDR hardware, processed, and sent to the computer, however, their frequency characteristics are changed. No matter which frequencies you originally configure your SDR to receive, the computer will always see zero-centered data. That's why the QT GUI Frequency Sink connected to the input data originally showed the frequency plot centered around 0 Hz, as you saw in Figure 6-3.

If you wanted, you could just agree with GNU Radio Companion and treat everything as zero-centered with respect to frequency. However, it's useful to be able to work with signals as if they occupy the frequencies that correspond to their actual physics. Working with signals in the frequency domain is more intuitive when your plots show those signals at their real-world frequencies. Without that, you'd have to constantly keep two sets of values in your head as you examine frequency plots and enter tuning values.

This is why there's a center_freq value in the AM receiver flowgraph. It allows you to have the best of both worlds: the real-world frequencies that make sense to you and the zero-centered frequencies that make sense to GNU Radio. All user-focused inputs and outputs (the QT GUI elements) use real-world frequencies. The tuning-related blocks (in this case, just the Signal Source) use the center_freq value to provide an offset, converting the real-world numbers to zero-centered numbers. Specifically, we accomplish this by setting the frequency of the Signal Source using the expression center_freq - freq. When you want to tune to 880 kHz, this expression will evaluate to 900e3 - 880e3 = 20e3. This has the net result of shifting the radio data 20 kHz to the right, centering the desired signal around 900 kHz in real-world terms, or around 0 Hz in GNU Radio terms.

NOTE   *If you want to see what the AM receiver flowgraph looks like without all this double bookkeeping, change the Default Value of* center_freq *to* 0 *and the Default Value of* freq *to* 20e3. *Everything will function just as it did before, but all the user inputs and displays will see the same zero-centered world that GNU Radio sees.*

This idea of multiple simultaneous frequency values (the real-world frequency and the GNU Radio frequency) can be difficult to get used to. The more you work with radio flowgraphs, however, the clearer it will become.

Tuning the AM Receiver

We're now ready to return to the AM radio flowgraph to see how the tuning process works in context. We'll add two FFTs to the flowgraph so you can see the results of each stage of the tuning. Add the first QT GUI Frequency Sink to the output of the Multiply block, with a Name of Shifted, and the second to the Low Pass Filter output, with a Name of Filter Out. When you're done, you should have something like Figure 6-23.

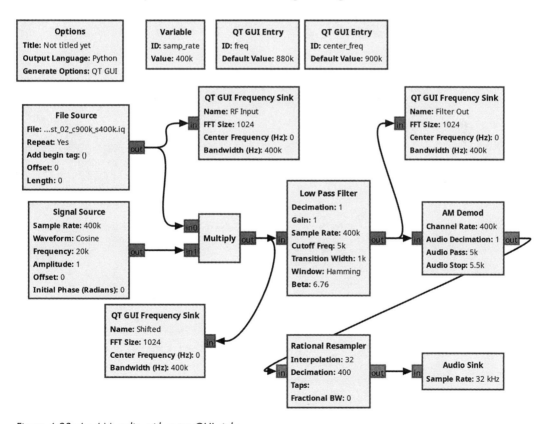

Figure 6-23: An AM radio with extra GUI sinks

Executing the flowgraph shows you how each stage of the tuner works. Figure 6-24 shows the execution window.

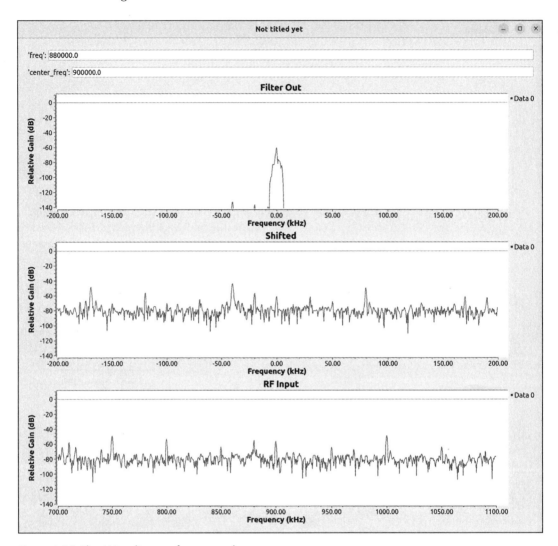

Figure 6-24: The AM radio tuner frequency plots

On the RF Input plot, you see the captured SDR data coming from the file, realistically centered around 900 kHz. Notice the peak we want to tune to, at 880 kHz, just to the left of the center of the plot. In the Shifted plot, that peak has been shifted to the center of the frequency axis. This second plot's horizontal axis is centered around 0 Hz, since you didn't change the plot's Center Frequency (Hz) property. Once you've gone through the frequency-shifting phase of tuning, it makes sense to start thinking of your data as zero-centered. You're finally on the same page as your software!

Finally, the Filter Out plot shows the effects of low-pass filtering your centered RF data. You can see the peak and a 5 kHz range on either side (remember the cutoff frequency?), but all the rest of the RF data is gone. This includes all the other signal peaks, as well as the frequencies between them. Failing to filter out those other signal peaks is the aural equivalent of listening to many people talking at the same time. We want our radio receiver to listen only for a single voice. And as you'll see in Chapter 8, the RF energy between those signal peaks is just *noise*. We don't want that either.

Try tuning to a few of the other signals you can see in the Input RF flowgraph and watch how the tuner operates. Then congratulate yourself on learning the art of tuning, one of the most significant concepts in radio communications.

Demodulation

Once you tune your AM receiver to a particular signal, the next step is to demodulate the signal. Recall that we discussed amplitude modulation and demodulation in Chapter 1. Specifically, we looked at a simple example where a faster (higher-frequency) carrier wave was modulated based on a slower (lower-frequency) sinusoidal signal. It looked something like Figure 6-25.

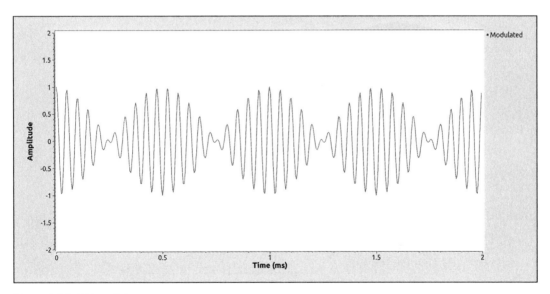

Figure 6-25: AM modulation

In this time-domain plot, we can see that the waveform fluctuates up and down in evenly spaced intervals, according to the frequency of the carrier wave. But instead of each peak having the same height, the height varies from peak to peak, tracing the shape of the lower-frequency sinusoid. This is the effect of amplitude modulation: the amplitude (height) of the carrier wave is changed based on the current level of the signal we wish to transmit.

Returning to your flowgraph, once the data emerges from the low-pass filter, you have a single AM radio signal. All the other signals and noise have been filtered out. Your AM signal consists of a carrier wave at a certain constant frequency that has been modulated in the same way as the simple example in Figure 6-25. Instead of changing the carrier wave's amplitude based on a simple sinusoid, however, the amplitude was changed based on a more elaborate audio waveform generated by a microphone into which someone was talking.

Demodulating this signal manually would require a fairly complicated design, with scary things called "envelope detectors" or "product detectors." Fortunately, GNU Radio has a simple block that just takes care of all that for you: the AM Demod block. This block takes a single AM modulated signal (which you have) and outputs the audio waveform it contains (which you want). The AM Demod is the next block in your *second_am_rx.grc* flowgraph, appearing right after the Low Pass Filter.

Taking a close look at the flowgraph, you can see that a signal of Complex type goes into the AM Demod from the Low Pass Filter, while a signal of Float type comes out. You can tell this from the color of the input/output tabs: the AM Demod has a blue input tab (Complex) and an orange output tab (Float). You can actually divide up the entire flowgraph into two parts hinging around the AM Demod: the Complex portion on the input side and the Float portion on the output side (see Figure 6-26).

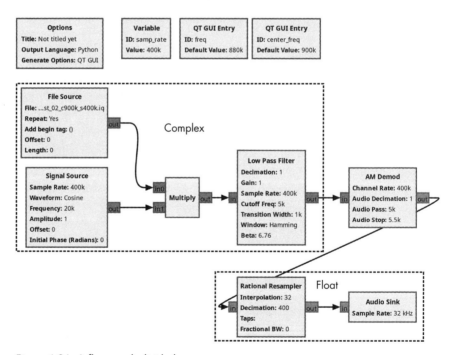

Figure 6-26: A flowgraph divided

As mentioned earlier, radio data tends to be complex. We also mentioned that audio data will be composed of real numbers (numbers with decimal points that aren't complex), which are typically represented in computers by floating-point types. Often demodulators will serve as the dividing line between the "radio" side of things and the "rest-of-the-world output" side of things, which in this case is our audio. As the data passes through the demodulator, its type is converted from Complex to Float.

Viewing the Modulated and Demodulated Signals

Now that we know what the AM Demod block is doing, let's use some instrumentation blocks to look at the data going into and out of the block. You already have a QT GUI Frequency Sink connected to the output of the Low Pass Filter. Add another one to your flowgraph, connect it to the AM Demod block output, set its Name to Demod Out, and set its Type property to Float. Before running the flowgraph, you might also want to disable the RF Input and Shifted GUI blocks; you don't need them anymore. When you're done, you'll have something like Figure 6-27.

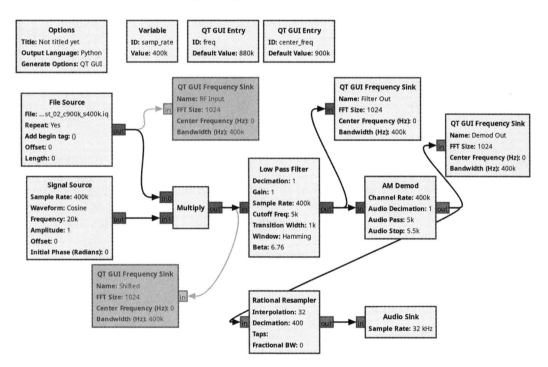

Figure 6-27: A flowgraph to display the AM demodulator output

Think about what this new frequency plot will show you when you run the flowgraph. With your first glance at the execution window in Figure 6-28, you might think, "Not much!"

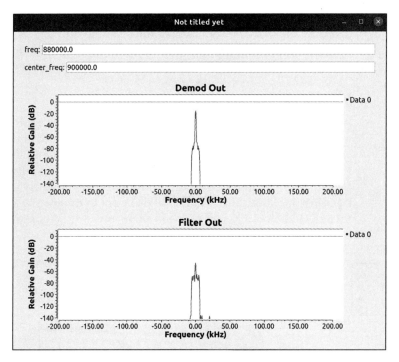

Figure 6-28: The AM demodulator output

The main thing you should notice is how the signal bounces up and down periodically. This is because the audio will get louder and softer at times, and this intensity is reflected in both the modulated and demodulated versions of the signal. This is a characteristic of amplitude modulation: larger audio signal, larger carrier; smaller audio signal, smaller carrier.

Setting the AM Demod Block Properties

The AM Demod block has several properties that we set when we built the AM receiver flowgraph in Chapter 4. We'll look briefly at the meaning of those properties now:

Channel rate   This is simply another name for the sample rate. We set it to 400k using the samp_rate variable.

Audio decimation   We'll look at decimation in the next section. For now, understand that setting this property to 1 means it will have no effect.

Audio pass   This is another term for *cutoff frequency*. The AM Demod block contains its own low-pass filter to get rid of any noise that may be part of the input or that might be generated by the demodulation process. This property sets the cutoff frequency for that internal low-pass filter. The default value for this property was 5e3, or 5 kHz. This is set to work with standard broadcast AM signals, which are mandated by government regulators to behave in specific ways. Unless you're building a custom AM transmitter and receiver that behave differently than standard AM broadcasts, you'll never need to change this.

Audio stop   This is another way of describing the transition width of the AM Demod block's low-pass filter. Instead of defining the size of the transition zone, the audio stop just defines where it ends (in this case, 5,500 Hz). Arithmetically, the audio stop minus the audio pass is equal to the transition width. This is illustrated in Figure 6-29.

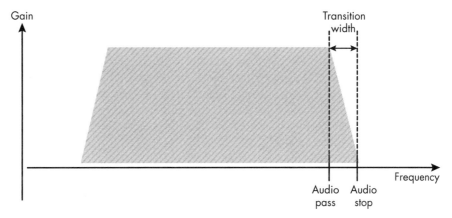

Figure 6-29: The audio pass and audio stop of an AM Demod block

We'll get into other types of modulation and demodulation besides AM in a later chapter. For now, let's be grateful for how powerful these GNU Radio blocks are and move on to the last stage in your AM radio.

Resampling

Now that you've tuned to an AM radio signal and demodulated it to produce an audio signal, the last step in your flowgraph is to resample the audio so your computer's sound card can play it. *Resampling* means changing the sample rate of a signal. This step is necessary because the sample rate of the audio signal coming out of the AM Demod block is 400 kHz, as shown by the block's Channel Rate property, but 400 kHz is too fast for the audio cards in most computers. In fact, the Audio Sink at the end of the flowgraph has a default sample rate of 32 kHz.

Here's the first rule of sample rates in GNU Radio: *they need to match when going from one block to another.* Put another way, the input sample rate to any block must be the same as the output sample rate of the block to which it's connected. Thanks to this rule, we can't simply connect the AM Demod block to the Audio Sink. The sample rates don't match. We need to *resample* the data, meaning change the sample rate, by adding a resampler between the two blocks. This block won't change the underlying character of the data, but it will convert the sample rate from 400 kHz to 32 kHz.

There are two basic types of resampling: *decimation* and *interpolation*. Decimation reduces the sample rate, while interpolation increases it. In our case, we'll actually need a combination of both.

Decimation

Decimation converts from a higher to a lower sample rate by simply getting rid of some of the samples. To see how this works, think about the number of samples that are taken over a fixed period of time. For example, if your sample rate were 1 ksps (otherwise known as 1,000 samples per second), you would have 1,000 samples in a 1-second interval. To cut the sample rate in half, you could simply throw away every other sample, leaving 500 samples over that same 1-second time interval. Since 500 samples divided by 1 second yields 500 samples per second, you can see that you've halved the sample rate for that 1-second period. Now, rather than just doing this for a single second, imagine that you always throw away every other sample. If you do, then you can see how the sample rate would be halved, as shown in Figure 6-30.

Figure 6-30: Decimation

Cutting the sample rate in half is the same as decimating the signal by 2. As the figure illustrates, data enters the resampler as a stream of evenly spaced samples, and it leaves the resampler as a more widely spaced stream of samples, having been decimated by 2.

Let's create a simple project to see what this looks like in practice. Name this new project *decimation.grc*. One note before you begin: the Type for all blocks used in this flowgraph will be Float, so please configure each block as such. To save a bit of time, you can select a block and press the up arrow or down arrow key to cycle through block data types. To go from Complex to Float, simply press the down arrow once.

First, add a Signal Source and connect its output to a Throttle block. Other than the Type, use the default settings for both of these blocks. Then, feed the Throttle output into a Keep 1 in N block, changing the Type to Float and the N property to 4. This block will decimate the incoming signal by a factor of 4. Finally, add two QT GUI Time Sink blocks to view the test signal before and after decimation. Connect the first to the output of the Throttle block, setting the Name to 32 ksps Input and the Number of Points to 128. Then click the **Config** tab and set the Line 1 Marker to Circle. Connect the second time sink to the output of the Keep 1 in N block. Set the Name to Decimated Output, the Sample Rate to 8000, and the Number of Points to 32. As with the first sink, also set the Line 1 Marker to Circle. When you're done, your flowgraph should look like Figure 6-31. (Fiddling around with the number of points and the markers of the time sinks isn't necessary from a functional standpoint; it's solely to make the output in this learning exercise look clearer.)

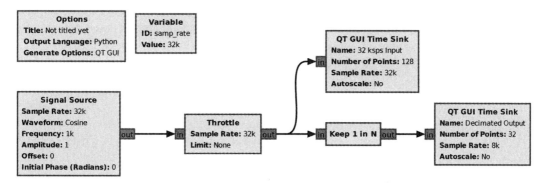

Figure 6-31: A decimation flowgraph

When you run the flowgraph, you'll see the image in Figure 6-32.

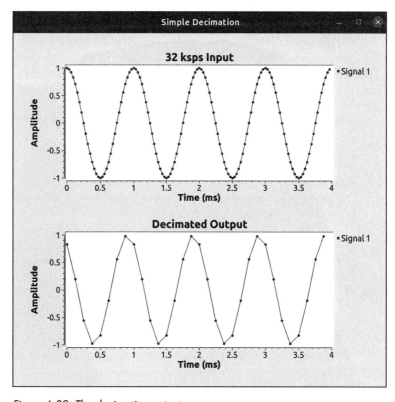

Figure 6-32: The decimation output

In these time sinks, each sample of the signal is represented by a dot, and GNU Radio helpfully draws a curve connecting these dots to show you the shape of the waveform. Notice how the Decimated Output has the same general shape as the 32 ksps Input, but it has fewer samples over the same time period. Specifically, it has one-fourth as many samples, since the decimation drops the sample rate from 32 ksps to 8 ksps (32,000 / 4 = 8,000).

You can decimate any signal this way, but be careful that you don't decimate too much, or you'll run afoul of the problem we encountered when we first discussed digital sampling in Chapter 2: distortion due to undersampling. We'll return to this issue in Chapter 11. For now, we'll assert that the sampling rate in our simple test project, 8,000 samples per second, is sufficient, since it's still significantly faster than the 1 kHz signal we're sampling.

Go ahead and try some different decimation values in the Keep 1 in N block to ensure it behaves as you expect. Note, however, that you can decimate only by an integer value. It makes sense to decimate by 7, for example, because you can keep every seventh sample and throw the rest away, reducing the effective sample rate by a factor of 7. The same method can't be used to decimate by 7.5, though; you can't keep 1 sample and throw away the next 6.5 without additional processing. As you try different decimation values, just remember that you'll have to adjust the Sample Rate and Number of Points of the Decimated Output time sink to produce the same rich visual aesthetic you saw here.

THE THROTTLE BLOCK

We've now used the Throttle block in a few flowgraphs without really explaining why. Let's take a moment to look at what it does. In short, it keeps your computer from working harder than necessary.

If you tried building some of the previous flowgraphs without a Throttle, they would most likely seem to work fine. You might sense some sluggishness from your computer, but otherwise it probably wouldn't have seemed much different. If you brought up your System Monitor, however, you would have seen at least one of your CPUs working as hard as it could.

It turns out that the sample rates you specify in your GNU Radio flowgraphs aren't exactly what they seem. Sometimes, the computer is actually going much faster. For the 32 ksps rate in the *decimation.grc* flowgraph, for example, we would expect our blocks to complete the math for the flowgraph once every 31.25 microseconds (1 divided by 32 kHz). Modern computers, however, can perform billions of operations per second, so they can easily do all the flowgraph's calculations at a much faster rate than once every 31.25 microseconds.

What does the computer do once it's finished one cycle's worth of calculations early? When we run a flowgraph that contains an interface to the physical world, such as an SDR-interfacing block or an Audio Sink block, the computer heeds the sample rate and will sit around idling until it's time to process the next data sample. But when everything in the flowgraph is synthetic, as in *decimation .grc*, something a little strange happens. Since there's no SDR hardware or sound card that's expecting data at a specific point in time, GNU Radio doesn't really need to wait for anything. It's essentially just running a simulation, so why not run it as fast as possible, regardless of the sample rate? That's exactly what your computer does, greedily hogging resources you'd like to use for other operations, like moving your mouse around and clicking things.

The solution is to simply add a Throttle block somewhere in the flowgraph. It forces data to slow down to whatever rate the block specifies. That's all you need to do to keep your CPU safe and sane. We find it easiest to put the Throttle right after one of the source blocks, but theoretically, it could go anywhere. For example, in *decimation.grc*, the Throttle could instead come after the Keep 1 in N block.

Interpolation

Interpolation increases the sample rate by creating extra samples between the existing samples in a signal. Each extra sample is added by estimating what value the original signal would have had at that point in time. For example, if your signal had a value of 3 at 1 ms and 7 at 2 ms, an interpolation algorithm might estimate the value at 1.5 ms to be 5 (Figure 6-33).

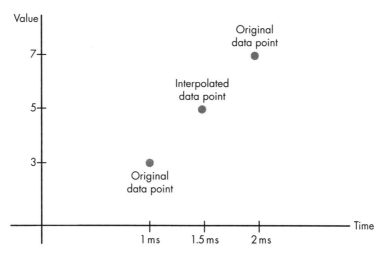

Figure 6-33: Interpolation

One warning about interpolation: you aren't actually capturing any new information about the sampled signal with interpolation. You haven't made any additional measurements of any real signals. You've just performed a mathematical trick that helps you match the sample rates between your different flowgraph blocks.

Resampling in the AM Receiver

Let's now return to your AM radio and see how we can resample the demodulated audio signal to be suitable for output. The audio signal starts with a sample rate of 400 kHz and needs to be converted to the typical sample rate of 32 kHz for your sound card. At first it may seem obvious that we need to decimate. But by how much? Dividing 400 by 32, we get 12.5, which is a problem. We just said a few paragraphs ago that we can't decimate by noninteger values.

Our solution is to use a `Rational Resampler` block. This block allows for *fractional resampling*, or resampling by a noninteger value. The block does this by combining decimation and interpolation to achieve the same result as a fractional decimation. An easy way to do the math is to use the old sample rate in kHz for the decimation value and the desired sample rate in kHz for the interpolation value. In our case, that's 400 for the decimation value and 32 for the interpolation value. This means that the rate of the incoming signal, sampled at 400 kHz, will be divided by 400 and multiplied by 32, leaving you with an output rate of 32 kHz; that's exactly what we need to send the signal on to the `Audio Sink` so that we can hear it!

There are other reasons to change the sampling rate within a project besides matching the rate required to output a signal to your sound card. A common reason is to reduce the computational load of your flowgraph. A slower sample rate means fewer numbers going through that part of the flowgraph. Fewer numbers mean fewer computations per second are required, thus reducing the CPU load on your computer. It's a good practice to decimate when you can for the sake of computational efficiency.

Conclusion

Pause for a moment to consider what you've accomplished to this point:

- You've installed GNU Radio, which is a great start.
- You've learned about basic radio concepts such as gain, frequency, filtering, and modulation, and you've seen them in action.
- You've built and debugged simple flowgraphs.
- You've gotten to know the basic parts of a software-defined radio by building and analyzing an AM receiver.

The AM radio receiver you built may not have been the most complicated radio system imaginable, but throughout the rest of this book you'll see how to build increasingly more advanced analog receivers using the same basic framework. Some of the pieces will be a bit different, but there will be more similarities than you might expect. Even if you switch from analog to digital receivers, the similarities won't end: those radios are structured with a lot of the same pieces as well. Because this framework is so crucial to understand, we'll continue your journey in the next chapter with a new kind of analog receiver.

7

BUILDING AN FM RADIO

In this chapter, you'll see how to build a different type of analog receiver: an FM broadcast receiver. Along the way, you'll get more practice working with GNU Radio. You'll meet some powerful new blocks and practice using variables and expressions to make your flowgraphs more flexible.

You'll develop the FM receiver in two stages. First, you'll make the absolute minimum of changes necessary to convert your AM receiver flowgraph from Chapter 4 into an FM receiver. The conversion will actually require very little modification to achieve, which illustrates the incredible modularity of software-defined radio; different SDR-based receivers have a lot in common with each other. After getting your basic FM flowgraph working, you'll then make a number of improvements to it. The result will be a basic but powerful SDR receiver framework that can form the basis for a multitude of analog (and digital!) SDR receivers. You'll be able to integrate hardware into this framework to run your receiver on live broadcasts, as we'll explore in Chapter 9. That's probably what many of you have been waiting for, and we're almost there.

The *FM* in the FM receiver you'll create in this chapter is short for *frequency modulation*. Like amplitude modulation, this is a technique for embedding the characteristics of one signal into another signal. We'll talk about the theory behind how FM modulation and demodulation work in Chapter 10. For now, let's dive into building the FM radio project.

Converting from AM to FM

Your FM radio receiver will start with a source of radio frequency (RF) data, which will be fed into tuning and filtering stages to isolate the signal that you want. This signal will then be demodulated to produce the originally transmitted signal. After some resampling, you'll then play the audio. These are exactly the same steps you took previously to make your AM receiver! The only fundamental difference between your FM radio and your AM radio will be the demodulation step. There will, however, be a couple of smaller tweaks along the way.

Start by opening your AM radio receiver flowgraph (or you can start with *ch_07/am_rx.grc*, included in your downloaded project bundle) and save it with a new filename. At first, the flowgraph should look like Figure 7-1. We'll only be swapping out the AM Demod block, while the rest of the flowgraph will just require some parameter changes.

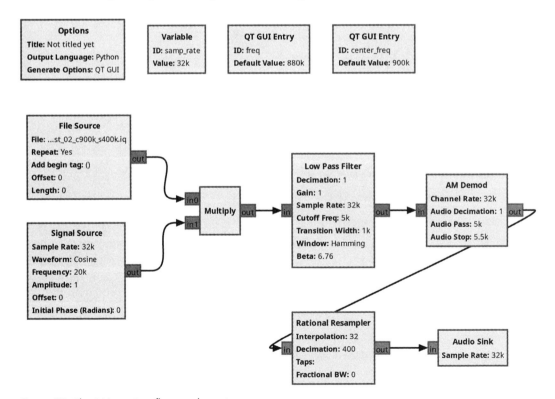

Figure 7-1: The AM receiver flowgraph again

You'll start making changes at the input side, beginning with the File Source and moving toward the output side. Open the File Source block and set the File property to *ch_07/fm_c96M_s8M.iq*. To properly process the RF data coming from this file, we need to know the center frequency and sample rate the SDR was operating at when the data was captured. The filename contains two clues: *c96M* denotes a center frequency of 96 MHz, while *s8M* tells us the SDR's sample rate was 8 Msps.

NOTE *You won't always have these parameters embedded in the filename, but you will need to know what they are somehow. Our advice: choose a filename with this type of format if you're creating it, or change the name of a file you're using if it doesn't have this information in it. You may think you won't lose that sticky note with these two parameters scribbled on it, but you shouldn't risk it.*

The AM receiver flowgraph had different values for these parameters, so let's update those now. Change the center_freq Default Value to 96e6 and the samp_rate Value to 8e6. Also, go ahead and set the Default Value of freq to 94.9e6. There are a number of signals in this file, and 94.9 MHz is one of them, so this change gives you a head start on the tuning process. When you're done, your flowgraph should look like Figure 7-2.

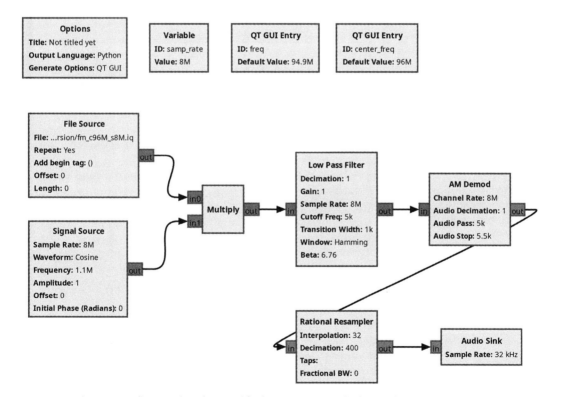

Figure 7-2: The receiver flowgraph with a modified File Source and other updates

Notice that the workspace's visual rendering of the Signal Source block has updated to show a Frequency of 1.1M based on the freq and center_freq values you've entered. This happens automatically thanks to the way you set up the Signal Source block with a variable as the frequency parameter; there's no need to change anything in the Signal Source to support the new freq and center_freq values. As we discussed in Chapter 6, the way to think of this 1.1M is that the file was originally captured with the SDR centered on 96 MHz, but after capture, the data is stripped of its original frequency information and will be interpreted by GNU Radio as if centered at 0 Hz. In that zero-centered universe, the broadcast signal originally at 94.9 MHz is now at −1.1 MHz. Multiplying the RF data by a complex sinusoid with frequency 1.1 MHz will shift the target signal to 0 Hz, otherwise known as the center. A tuning scenario with these values is shown in Figure 7-3, with both sets of frequencies labeled.

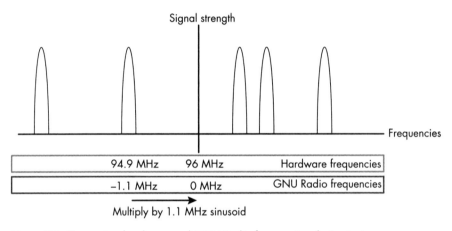

Figure 7-3: Comparing hardware and GNU Radio frequencies during tuning

The next block to look at is the Low Pass Filter. Recall that once you've centered your target signal, you want to filter out all but a limited range of frequencies centered around zero. We set the cutoff frequency of this filter based on the width of frequencies we expect the target signal to occupy. Later we'll define this quantity more carefully and call it *bandwidth*, but for now, take it on faith that a conservative estimate for the width of an FM broadcast signal is 150 kHz.

In light of this, you might be tempted to set the Cutoff Freq of the Low Pass Filter to 150e3. Recall, however, that this complex low-pass filter will pass frequencies between the negative cutoff frequency and the positive cutoff frequency. If you set the cutoff to 150e3, you'll pass the frequencies from −150 kHz up to +150 kHz, for a total bandwidth of 300 kHz. This is too wide and will have the effect of letting through a bunch of extra noise, and possibly other signals, that you should be filtering out. Your system might work with this setup, especially if the signal you want is very loud and clear and if there are no other signals close by in frequency. Still, we wouldn't recommend unnecessarily letting this extra noise into your system. Instead, set the value of the Cutoff Freq to 75e3, which will result in a 150 kHz range of frequencies passing through the filter. Also change the Transition Width to 7.5e3 (our one-tenth rule of thumb). Finally, change the Decimation to 20.

This last change will simplify the rest of the flowgraph a bit. The original flowgraph ran at a sample rate of 400 ksps right up until the end, when it dropped down to 32 ksps so that the Audio Sink would work properly. Since the first blocks of your new flowgraph are running much faster than before, we'll have to account for this difference at some point. Dropping the sample rate of the data exiting the Low Pass Filter down to 400 ksps will allow the latter part of the flowgraph to run at the same sample rate as before, minimizing the changes you'll need to make to the subsequent blocks. Why a value of 20? Because 8 Msps, or 8,000 ksps (the sample rate of the .iq file), divided by 20 is 400 ksps. When you complete your property changes, your flowgraph should look like Figure 7-4.

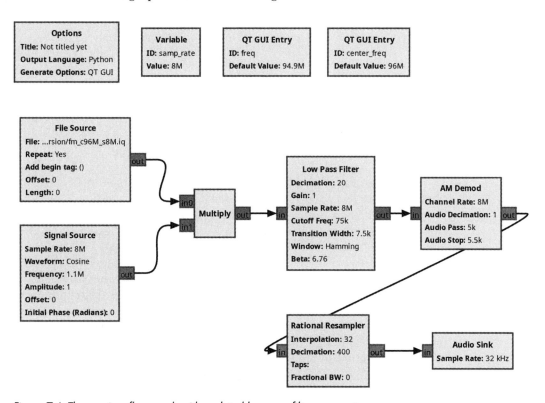

Figure 7-4: The receiver flowgraph with updated low-pass filter parameters

Next up is the demodulator step. You'll need an entirely new block for this stage, so delete the AM Demod and add a WBFM Receive block in its place, connecting its inputs and outputs in the same way. Set the block's Quadrature Rate to 400e3 (this is just another name for the sample rate) and the Audio Decimation to 1 (we don't want to change the sample rate any further). When you're done, your flowgraph should look like Figure 7-5.

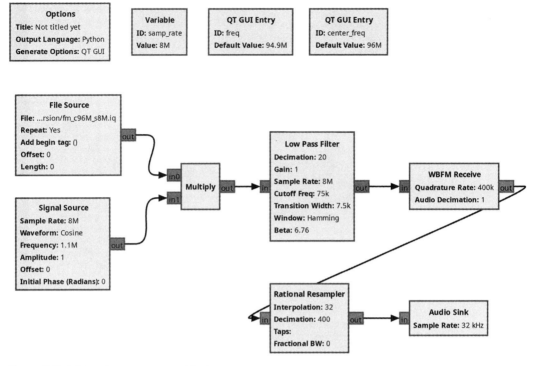

Figure 7-5: Adding the WBFM Receive block

The beginning of the flowgraph is running at a sample rate of 8 Msps, but the Rational Resampler is still being fed a demodulated audio signal sampled at 400 ksps, just as before. Also as before, you're sending data to the Audio Sink at 32 ksps. This means we don't need to change anything else. Your flowgraph should work with no more changes required!

Before running the flowgraph, check your computer's audio levels and make sure they're below 50 percent. This flowgraph drives the Audio Sink with a much stronger signal than the AM receiver, and there's no point in waking up the neighborhood. Also, driving your audio card with too strong a signal can result in unpleasant distortion. When you're at a reasonable level, execute the flowgraph. You should hear a voice talking almost immediately. The audio will repeat very quickly because the captured file is only a few seconds in duration. As we'll see later, captured RF files can get very large, so we made this one small enough not to be unwieldy.

At this point, you have a functional FM receiver—congratulations! As you've seen, it isn't all that different from your original AM receiver. The key takeaway is that SDR-based receivers share a great deal in common with each other. If you're already looking ahead to digital radios, a lot of what we do in the next book will depend on this basic receiver structure.

Improving the FM Receiver

You've gotten your FM receiver working, but you're not done. There are several things you can do to make the flowgraph better, and we're going to start on that now. The final FM receiver flowgraph is going to be functionally equivalent to what you already have, but it will have these key enhancements:

- A new block to do all the tuning steps in one place
- A cleaner and more powerful set of variable definitions
- A graphical slider for volume control

We'll also add an instrumentation block so you can easily view the RF data and tune to different frequencies.

Tuning More Effectively

The current tuning implementation works, but there's a cleaner way. Right now, it takes three different blocks (the Signal Source, the Multiply, and the Low Pass Filter) to tune the signal, but GNU Radio has a single block that combines all this functionality. It's called the Frequency Xlating FIR Filter block—quite a mouthful. This block will perform all the same tasks as the existing tuner: frequency shifting, filtering, and decimation.

Start by deleting your existing tuner, removing the Signal Source, the Multiply, and the Low Pass Filter blocks. Then add a Frequency Xlating FIR Filter to your flowgraph, connecting its input to the File Source and its output to the WBFM Receive. As you add this block, be careful not to mistakenly add the Frequency Xlating FFT Filter instead. This other block will sometimes work, but it's typically much more computationally intensive. When you're done, the flowgraph should look like Figure 7-6.

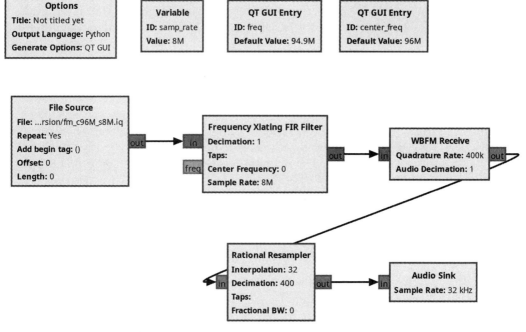

Figure 7-6: The FM receiver flowgraph with a `Frequency Xlating FIR Filter` *block*

The technique you used in your original tuner to frequency-shift the RF data has several different names. In the hardware world, we'd call it *complex mixing*. If we just look at the flowgraph math, we might term it *complex sinusoidal multiplication*. Another description, though, is *frequency translation*. Hence, the first part of the new block's name tells us that it translates, or shifts, the frequency of the input data. *Xlating* is just a slight abbreviation of *translating*.

The second half of the name refers to a type of filter. A *finite impulse response (FIR) filter* is a specific way of implementing a filter on sampled data. The theory behind these FIR filters is beyond the scope of this book; you can find a lot of great resources online if you're interested. For tuning purposes, we'll be building a low-pass filter using an FIR scheme, but you don't need to know how FIR filters work to make that happen—GNU Radio does the work for you.

We need to set three properties on the `Frequency Xlating FIR Filter` block. The first is the Center Frequency, which controls how to shift the frequency of the input RF data. Its name can be a little confusing at first because we already have something called `center_freq` in our flowgraph. It's natural to assume these are the same things, but they aren't. As we proceed, you'll see how they differ.

GNU Radio assumes that the input data is centered around 0 Hz. The Center Frequency property directs the block to recenter the RF data around a different frequency. Let's illustrate this using the same example shown earlier in Figure 7-3. If we use −1.1 MHz for the Center Frequency value, the RF data will shift to the right such that the –1.1 MHz point will now be in the center. Figure 7-7 shows this example with both the GNU Radio frequencies and the hardware frequencies.

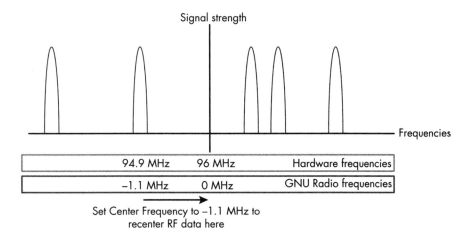

Figure 7-7: Tuning with the Frequency Xlating FIR Filter

This is similar to how we did frequency shifting before, but the shift quantity now has the opposite sign. Using the original, three-block method on our example, we would have multiplied the RF data by a complex sinusoid with frequency of +1.1 MHz to center on the target signal, rather than selecting –1.1 MHz for the Center Frequency property. With the old method, we specified the amount of frequency to shift (+1.1 MHz). The new method requires us to select the frequency to center on (–1.1 MHz). The sign on the value is the only difference, as shown in Figure 7-8.

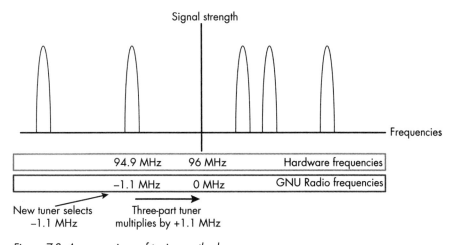

Figure 7-8: A comparison of tuning methods

There's one additional complication, however: although the computer sees flowgraph data as centered at 0 Hz, we humans generally prefer to think in terms of real-world frequencies. In other words, we should present the real-world frequencies to the flowgraph's human user, while doing arithmetic so the flowgraph has the zero-centered values it needs to work. This means we need to compute the Center Frequency property for the tuner using an expression of the hardware frequency variables: freq and center_freq. Recall that this center_freq value denotes the frequency at which the SDR hardware was originally configured to capture the RF data we have in the file. The freq variable represents the actual hardware frequency to which we'd like to tune.

With this in mind, double-click the **Frequency Xlating FIR Filter** block and set the Center Frequency to freq - center_freq. Then click **OK**. Notice that this is simply the opposite of what we did in the three-block approach, where we set the frequency of the Signal Source block to center_freq - freq. Switching the variables in the expression gives us a value with the correct sign.

The next tuner function we need to configure is the low-pass filter. Before entering anything for that, you should create two more variables. First, copy and paste the samp_rate Variable block, and change the ID of this copy to chan_width and the Value to 150e3. This channel width variable denotes the range of frequencies we want to pass through the filter. It's not quite the same as the Cutoff Freq property on the old Low Pass Filter block; the filter passes frequencies between the negative cutoff and the positive cutoff, whereas the channel width describes the entire range of frequencies, both positive and negative. The channel width is therefore twice as large as the cutoff frequency, as shown in Figure 7-9. We use this value rather than the cutoff frequency because it's a bit more intuitive to think in terms of widths, rather than half-widths.

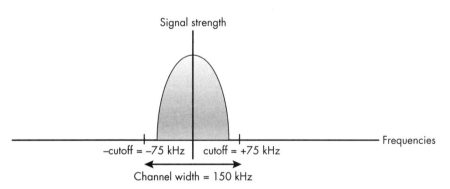

Figure 7-9: Cutoff frequency versus channel width

Next, copy and paste the chan_width variable to create another duplicate. Change the ID of the duplicate to transition_width and the value to chan_width/20. This is simply applying our rule of thumb that a decent transition width is one-tenth the cutoff frequency. Since the channel width is twice the cutoff frequency, we divide by 20 instead of 10.

Now to make use of our two new variables. Without going into pages of digital filter theory, we will just tell you that the function of an FIR filter is driven by a set of mathematical quantities called *taps*. We definitely do not want to figure out how to calculate these taps by hand, so bear with us as we invoke a little bit of magic. Double-click **Frequency Xlating FIR Filter** and enter the following into the Taps property field:

```
firdes.low_pass(1, samp_rate, chan_width/2, transition_ width)
```

You've just done some Python programming! All this does is call a GNU Radio function that generates the taps for a low-pass filter. You're passing the function four arguments in the parentheses. The first is the filter gain. We intend only to filter, not apply any gain, so we've set this to 1. The second argument is the sample rate of the data entering the block. This is the samp_rate variable you've already defined. The third argument is the cutoff frequency of the filter. As you saw in Figure 7-9, the channel width is twice the cutoff frequency. Conversely, the cutoff frequency is half the channel width, so we've entered chan_width/2 here. Finally, the last argument is the transition width. We created a variable earlier for exactly this purpose, which we've entered here as transition_width.

After the Taps field, the last property to set is the Decimation. Change this to 20 to match the previous decimation, which reduced the 8 Msps rate at the block's input down to the 400 ksps output rate.

Now that you've replaced the tuner and updated the Frequency Xlating FIR Filter properties, your flowgraph should look like Figure 7-10.

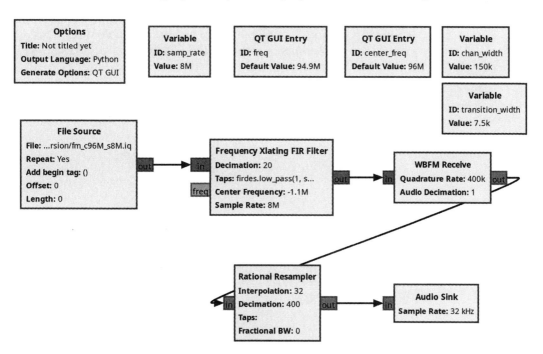

Figure 7-10: The FM receiver with a configured Frequency Xlating FIR Filter property

Execute the flowgraph to ensure that it still works. Even though this is a much more compact tuner, it should function identically to the more cumbersome method we used previously.

Updating Variables Automatically

Next, we'll expand our use of variables to make the flowgraph more flexible. Everything currently works, but what if you decided you wanted to decimate down to a different sample rate than 400 ksps? You'd have to update the properties of three different blocks to implement this change. Tweaking the behavior of your flowgraphs will be common as you debug and optimize their performance. The process is far more efficient if you can make a change by updating the value of a single variable, rather than adjusting multiple blocks.

In this case, the key is to think not in terms of decimation, but in terms of the reduced sample rate that we actually want exiting the Frequency Xlating FIR Filter. We'll create a variable for that purpose. Copy and paste the transition_width variable, change the new variable's ID to working_samp _rate, and set its Value to 400e3. We'll use this new quantity several times throughout the flowgraph to control the sample rates.

First, change the Decimation property of the Frequency Xlating FIR Filter to int(samp_rate/working_samp_rate). Rather than hardcode the decimation, this expression computes it based on the starting sample rate of the flowgraph (samp_rate) and the sample rate desired downstream of the tuner (working_samp_rate). Recall that decimation can be done only by integer values. Because any Python division operation will result in a floating-point value, we use the int() syntax to cast it, or force Python to consider it as an integer. The block properties should now look like Figure 7-11.

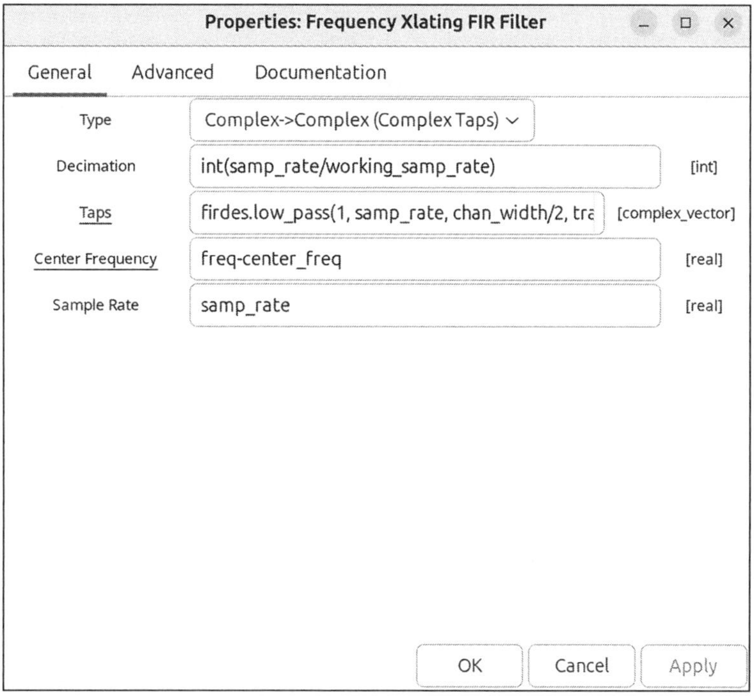

Figure 7-11: The Frequency Xlating FIR Filter *property window*

The Frequency Xlating FIR Filter now automatically computes its decimation, but the downstream blocks still have hardcoded sample rates. To fix this, change the Quadrature Rate of the WBFM Receive block to working_samp _rate. Then double-click the Rational Resampler and change the Decimation to int(working_samp_rate/1000). This will produce an integer-valued decimation of 400, which in conjunction with the Interpolation of 32 will produce a 32 ksps output data stream. After clicking OK, the flowgraph won't look any different from Figure 7-10, except for the new variable. The difference is that changing a single variable in your flowgraph will now cause the properties of three different blocks to automatically update to their necessary values.

Execute your flowgraph again to ensure that nothing has broken. Then change the value of `working_samp_rate` to `200e3` and note how the entire flowgraph immediately updates to account for this new value. As shown in Figure 7-12, both Decimation properties and the Quadrature Rate have all updated to reflect the modification. Execute the flowgraph yet again to prove to yourself that what you've done works as well as we're claiming it will.

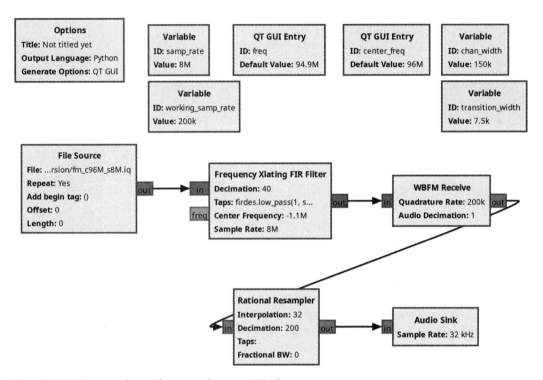

Figure 7-12: Changing the working sample rate to 200 ksps

You can change `working_samp_rate` back to `400e3` before continuing if you'd like your flowgraph to match exactly the screenshots for the rest of this chapter. If you want to leave it at `200e3`, that's okay, too. Well-designed flowgraphs don't only work with a single, magical set of values. They should work for any values that make sense.

Controlling the Volume

The last enhancement we'll make to the flowgraph is adding a real-time volume control. Up to now, you've used QT GUI Entry blocks to control your flowgraphs while they're running, but GNU Radio Companion provides several other GUI widgets that control variable values in real time. Another common widget is the QT GUI Range, which uses a slider graphic to control the value of the underlying variable. We'll use one of these widgets to adjust the output volume.

Add a QT GUI Range to the flowgraph and set its properties as follows:

ID volume

Type float

Default Value 5

Start 0

Stop 11

Step 1

Figure 7-13 shows the complete property settings for the block.

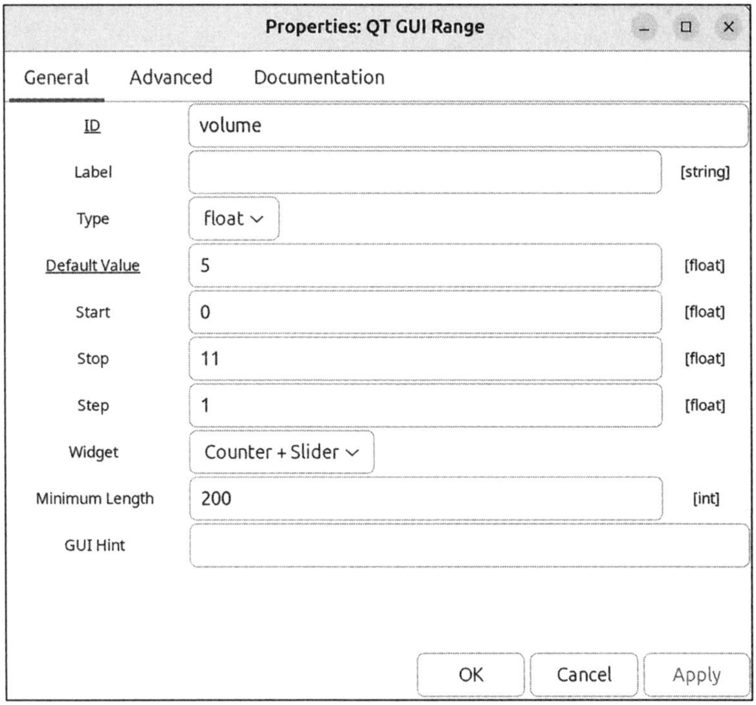

Figure 7-13: The QT GUI Range property window

You now have a variable called volume that you can use in your flowgraph. It will have a default value of 5 when the flowgraph starts, but it will provide a slider interface allowing you to select any value between the start (0) and stop (11) points. Because the step size is 1, you can select any whole number between those points. (Alternatively, if the step size had been 2, the only selectable values would be 0, 2, 4, 6, 8, and 10.)

To integrate the volume control into the flowgraph, break the connection between the Rational Resampler and the Audio Sink. Then add a Multiply Const block, setting the IO Type to float and the Constant property to volume/10. Finally, connect its input to the Rational Resampler output and its output to the Audio Sink input, as shown in Figure 7-14.

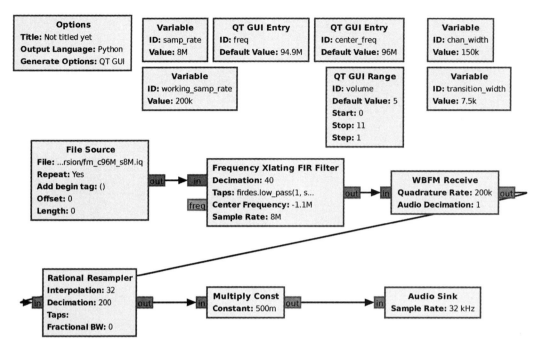

Figure 7-14: The FM receiver with volume control

Run the flowgraph. You should hear the voice again, but this time the sound level will be a bit lower. That's because you're attenuating, or applying gain less than 1, to the audio signal before outputting it to your sound card. Specifically, the attenuation is five-tenths, or 0.5—half of the original volume. You should also see a graphical volume slider, as shown in Figure 7-15.

Figure 7-15: The QT GUI *Range slider in action*

Drag the volume slider back and forth and note the change in sound level. Moving it to 10 should result in the same audio level you had before adding the slider. Moving it to 0 should mute the audio entirely.

Tuning to Other Signals

At this point, you've tuned only to a single frequency in the RF data, but there are many more. Let's add an instrumentation block to help find other signals in the data. Add a QT GUI Frequency Sink, change its Center Frequency (Hz) property to center_freq, and connect it to the File Source. Then press the right arrow key and notice that the block rotates 90 degrees clockwise. Press the right arrow key again to rotate another 90 degrees, effectively switching its input port to the opposite side. This doesn't change the block's function in any way, but having the block's input port on the right instead of the left can produce a cleaner-looking flowgraph, as shown in Figure 7-16.

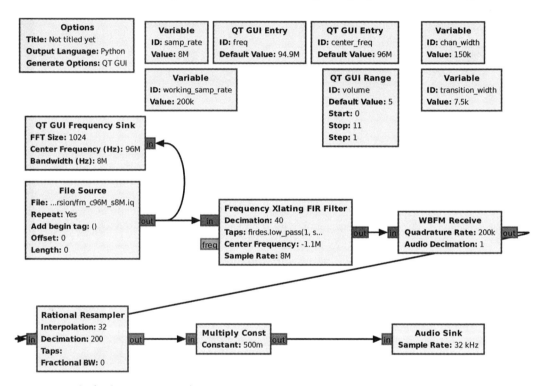

Figure 7-16: The final FM receiver with a QT GUI Frequency Sink

Execute the flowgraph again, and you'll now see a frequency plot of the RF data. Search the plot for signal peaks other than the default tuning of 94.9 MHz. As shown in Figure 7-17, you'll find signals at 92.5 MHz, 95.7 MHz, 98.1 MHz, and more. Notice that these frequencies correspond to stations you'd typically tune to on an FM radio, where they might be referred to simply as 95.2, 95.7, or 98.1.

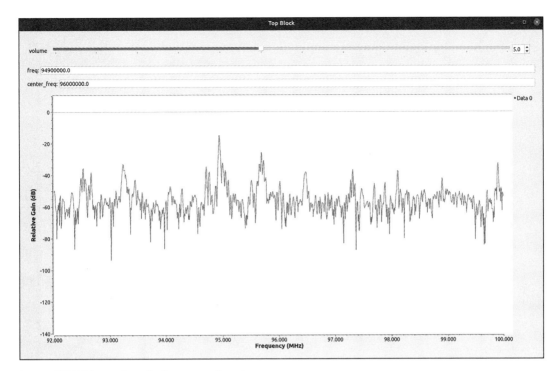

Figure 7-17: FM signals in the frequency domain

Take note of the frequencies corresponding to the peaks in the plot. Then try tuning to some of them using the QT GUI Entry that controls the freq variable. Enjoy exploring this FM radio data!

Conclusion

From one perspective, we didn't make huge changes to your AM receiver in this chapter. We simply converted your flowgraph to an FM receiver and then optimized its design. Looking at it another way, however, you've learned some extraordinary things. You've seen how to use more sophisticated GNU Radio blocks to make your flowgraph more efficient, and you've expanded your use of variables to make the design more generalizable. In the process, you've gotten a glimpse of the incredible flexibility of software-defined radio—it took only a few tweaks to go from AM to FM.

The core of your final receiver flowgraph, consisting of a Frequency Xlating FIR Filter and a demodulator, provides a framework for numerous analog and digital receivers. The clean and powerful management of variables in this final flowgraph will also make your projects much easier to modify and debug, saving you many hours of grief during your SDR endeavors. This will help in the coming chapters as we transition from software-only projects to projects that incorporate SDR hardware.

PART III

WORKING WITH SDR HARDWARE

8

THE PHYSICS OF RADIO SIGNALS

In this chapter, we'll go into more detail about radio signals: what they are, how they travel, how much "space" they take up, and the noise that makes it harder for them to do their job. Earlier chapters played somewhat fast and loose with these ideas on the theory that it was better to get off the ground quickly by learning with actual GNU Radio projects rather than spend hundreds of pages plowing through dry mathematical and engineering definitions. Now we're ready to tighten up some of these concepts.

Before proceeding, we want to invoke the onion model again. We'll still only be going so deep into these definitions, and they'll rarely be presented with the same mathematical formality of an engineering or physics textbook. By clearing up these concepts in a more general way, the goal is for you to more easily be able to make sense of radio-related materials you find online or in print, as well as converse with others on radio frequency (RF) topics. Remember, the word "fun" is part of "fundamentals" for a reason: once you understand these fundamentals, it's much easier to have fun with SDRs. Fortunately, you now know your way around GNU Radio, so you can work directly with the concepts behind these definitions, rather than just have us recite them to you.

Electromagnetic Waves

In Chapter 1 you learned that a signal consists of a physical property changing in a way such that it conveys information. The changing physical properties that make up radio signals belong to *electromagnetic waves*. Engineers categorize electromagnetic waves based on their frequency, arranging them along the *electromagnetic spectrum*. We focus on frequency as the defining property because it determines how a wave interacts with the environment, including how well it will penetrate structures; whether various layers of the atmosphere will absorb it, reflect it, or bend it; and even whether it will be visible to the human eye. The categories of the electromagnetic spectrum are shown in Figure 8-1, lined up from the lowest frequencies on the left to the highest on the right.

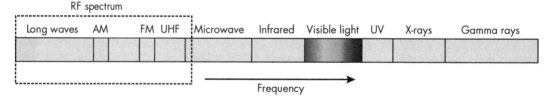

Figure 8-1: The electromagnetic spectrum, including the RF spectrum

You may recognize some of these groupings as radio related, while others are clearly not. For example, electromagnetic waves with frequencies roughly between 430 THz and 790 THz (*terahertz* are just 1,000 GHz, or 1 trillion hertz) are what you know as visible light. Unless you possess special mutant powers, everything you're seeing right now is brought to you by electromagnetic waves within that frequency range. Likewise, waves with frequencies between 3×10^{16} Hz and 3×10^{19} Hz are considered X-rays, the kind used in medical devices and airport luggage scanners. Go even higher than that and you'll find the kind of waves that can transform fictional experimental physicists into something very, very angry.

We're primarily concerned with the radio portion of the electromagnetic spectrum, which is highlighted on the left side of Figure 8-1. According to the Internal Telecommunications Union (ITU), a communications technologies group, the RF spectrum ranges from very low frequencies to 3 GHz. A common definition for the minimum radio frequency is 3 kHz, although some contend it should be 3 Hz. The question of how low is "very low" doesn't turn out to be very significant for most applications because radio waves are progressively harder to use at extremely low frequencies. As such, only specialized communications systems, like those in submarines or underground mines, use frequencies below 1 kHz.

Immediately above 3 GHz are what many call "microwaves." Here we have another issue with definitions, since lots of wireless communications systems that we think of as radio technology, including Wi-Fi, use these microwave frequencies. Also, satellite systems and 5G cellular communications systems can use something called *millimeter wave frequencies*, which are much, much higher than 3 GHz. For the purpose of this book, however, we'll limit ourselves to a subset of the radio spectrum: from around 1 MHz to 1 GHz. GNU Radio isn't limited to this range, but the SDR hardware you might want to use is another question. We'll explore the frequency limitations of SDR platforms in Chapter 11.

Keep in mind that the terms defining the different portions of the electromagnetic spectrum are largely human distinctions. Radio waves are labeled as such because they tend to travel in ways useful to wireless communication. There's no reason we can't build systems to send and receive communications on vastly higher frequencies than the RF spectrum, however. In fact, that's what fiber optic systems do; they modulate data onto a wave of light.

Propagation

Propagation describes how electromagnetic waves travel. When these waves encounter something, do they bend? Are they absorbed into the object? Are they reflected? The answers to these questions play a large role in determining whether a radio transmission will reach its intended recipient.

If radio transmissions were to occur only in a vacuum, with no obstacles of any kind between the transmitter and the receiver, then propagation would be much simpler to predict. In the real world, however, we have buildings, hills, valleys, weather, different atmospheric layers, the shape of the Earth, solar flares, meteor showers, and a host of other complicating factors. Radio waves of different frequencies interact with these factors in radically different ways. As a result, radio designers choose a frequency that best handles the factors expected to impact their specific case.

For example, radio waves at some frequencies can transmit to distant receivers by reflecting off the upper layers of the atmosphere and back down to Earth, bypassing a range limitation that would otherwise be imposed by the curvature of the planet. This is why you will sometimes pick up AM radio stations broadcasting from distant cities that might seem out of range. On the other hand, radio waves of other frequencies are much better when trying to communicate in dense urban areas. This is one of the reasons your cell phone communicates on very different frequencies than your car's radio.

Mathematically modeling the propagation of radio waves in real-world scenarios can be astoundingly difficult, but the general propagation characteristics of different ranges of frequencies are well known.

Frequency Bands

Radio engineers divide the RF spectrum into different *frequency bands*, ranges of frequencies that are best suited to particular applications. Table 8-1 provides a basic list of RF bands, including the common abbreviations and uses for each one.

Table 8-1: RF Bands

Frequencies	Bands	Uses
3–30 Hz 30–300 Hz 300 Hz to 3 kHz	ELF SLF ULF	Underwater communications, mineshaft communications
3–30 kHz	VLF	Atomic clock broadcasts, heart rate monitors
30–300 kHz	LF	AM longwave broadcasting, RFID, amateur radio
300 kHz to 3 MHz	MF	AM broadcast, amateur radio
3–30 MHz	HF	Shortwave broadcast, amateur radio, RFID, marine communications
30–300 MHz	VHF	FM broadcast, television, amateur radio, pagers
300 MHz to 3 GHz	UHF	Television, mobile phones, Wi-Fi, Bluetooth, GPS, amateur radio, pagers, home automation
3–30 GHz 30–300 GHz	SHF THF	Wi-Fi, amateur radio, satellite communications, mobile phones

There's a lot going on in the RF spectrum. For the purposes of this book, though, we're not going to concern ourselves with the extremely low and extremely high portions of the RF spectrum. This is because those bands typically require antennas that are extremely long (miles in some cases) or extremely specialized (like parabolic dishes). Additionally, the transmit and receive hardware often require some fairly exotic components not found in traditional software-defined radio hardware.

Many of the applications we're most interested in will be in the medium-frequency (MF), high-frequency (HF), very high-frequency (VHF), and ultra-high-frequency (UHF) bands. Speaking very generally, HF signals can have much longer ranges but only under the right environmental conditions. VHF signals typically propagate only along line-of-sight paths. UHF signals are also line-of-sight, but they're able to accommodate very high data rates when used for digital signals.

Again, these are gross generalizations, but they should help you see that different bands are best suited for different situations.

WAVELENGTH VS. FREQUENCY

Remember when we told you the frequency range of visible light was 430 to 790 THz? If you look this up in a physics textbook, you'll probably see the range referred to using a different characteristic, *wavelength*. This is the length of a single oscillation of a wave. Visible light has a range of wavelengths from 380 nm (nanometers, or billionths of a meter) to 700 nm.

Some parts of the radio spectrum are traditionally labeled by frequency and others by wavelength, so it's important to know the difference between these two properties. For an electromagnetic wave of a single frequency, the intensity of both its electric and magnetic components oscillates back and forth, as shown here:

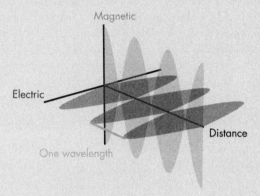

The oscillating pattern of an electromagnetic wave

This is our sinusoidal waveform, a consistently recurring entity in math and physics. The wavelength is simply the distance traveled by the wave between each cycle of the waveform, as shown in the figure.

Physics 101 tells us that the distance (*d*) traveled is equal to the speed (*v*) multiplied by the time (*t*):

$$d = vt$$

The wave will traverse a distance equal to a single wavelength (λ) in the time it takes the wave to oscillate once ($t_{oscillation}$):

$$\lambda = vt_{oscillation}$$

The time it takes to oscillate once is simply the reciprocal of the frequency (imagine flipping cycles per second and you get seconds per cycle):

$$\lambda = \frac{v}{f}$$

(continued)

The units for wavelength are then *distance per cycle* or, more commonly, just *distance*. Since electromagnetic waves travel through the air at the speed of light (please, physics purists, don't get hung up on the tiny difference between light's speed in a vacuum versus air), we get our formal equation for wavelength:

$$\lambda = \frac{c}{f}$$

Here λ is the wavelength, f is the frequency, and c is the speed of light, or 3×10^8 meters per second. Sometimes it's easier to think of the speed of light as 300 million meters per second, or 300×10^6. The equation is telling us that the wavelength of any electromagnetic wave is equal to the speed of light (in meters per second) divided by the number of oscillations it's making per second (in Hertz, or cycles per second). This yields the number of meters per cycle, which is another way to think about a wavelength. Here's an example using an FM radio signal at 100 MHz:

$$\lambda = \frac{300 \times 10^6 \text{ meters/sec}}{100 \times 10^6 \text{ cycles/sec}} = 3 \text{ meters/cycle}$$

You can also convert wavelength to frequency by reorganizing the equation as follows:

$$f = \frac{c}{\lambda}$$

If you needed to determine the frequency of a 1 m wavelength, you would therefore do it as follows:

$$f = \frac{300 \times 10^6 \text{ meters/sec}}{1 \text{ meter/cycle}} = 300 \times 10^6 \text{ cycles/sec} = 300 \text{ MHz}$$

Most of the time you won't need to do these calculations, but every once in a while you may run across an RF document that talks about wavelength rather than frequency. If you need to convert that wavelength to a frequency (or vice versa), there's your math.

Bandwidth

If the RF spectrum is divided into *bands,* the term *bandwidth* refers to the span of a range of frequencies. In particular, bandwidth describes how large of a frequency range a real-world radio signal occupies. Put another way, the bandwidth specifies how much of the limited space in the RF spectrum your signal takes up. In our FM receiver flowgraph from Chapter 7, bandwidth first enters the picture in the tuner, implemented by the Frequency Xlating FIR Filter block highlighted in Figure 8-2.

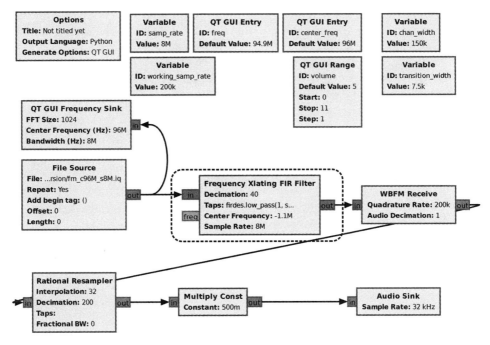

Figure 8-2: An FM receiver with the tuner highlighted

The filter has a cutoff frequency of 75 kHz, and because this filter is operating on a complex number stream, it's actually filtering from –75 kHz to +75 kHz, as shown in Figure 8-3. This 150 kHz filter passband is chosen to match the FM signal's expected bandwidth, and it's controlled by the chan_width value fed to the Frequency Xlating FIR Filter.

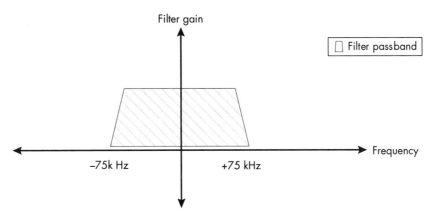

Figure 8-3: A passband for a complex low-pass filter

The FM signal you receive must "fit" inside the filter's passband, which we select based on the signal bandwidth (150 kHz, in this case) as shown in Figure 8-4(a). If the signal were unexpectedly wider than the filter's bandwidth, as in Figure 8-4(b), the filter would distort it by chopping off the higher-frequency components of the signal. This would degrade the performance of your receiver or cause it to fail altogether.

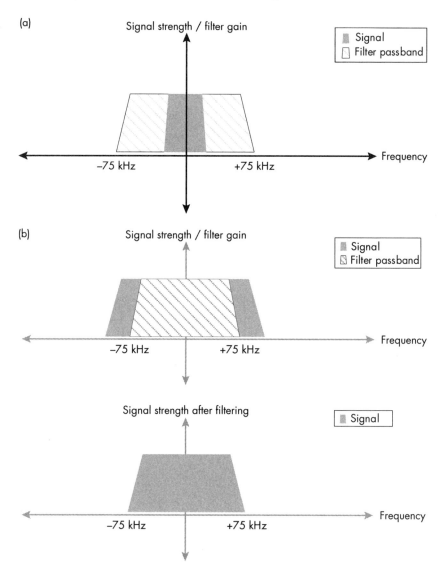

Figure 8-4: A signal and valid filter passband (a) compared to a signal and insufficiently large filter passband (b)

A closely related concept to bandwidth is *channel width*, the formal specification for how much of the RF spectrum a type of signal is authorized to consume. Consider the spacing on your FM radio dial. All the stations in the United States are located at odd multiples of 200 kHz: 94.7 MHz, 94.9 MHz, 95.1 MHz, and so on. Each station, or channel, is assigned a width of 200 kHz, as shown in Figure 8-5.

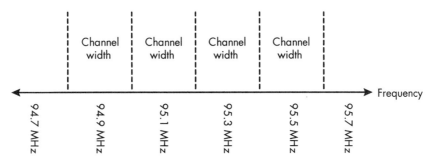

Figure 8-5: An FM channel map

Bandwidth is a physics issue; it's a physical property of the signal in question. Channel width, on the other hand, is a human (and often legal) construct, separate from the physics of actual RF signals. In theory, a signal is supposed to fit inside the channel defined for it. That is, the bandwidth should be no larger than the channel width, like in Figure 8-6(a). But what if, for example, an FM transmitter is poorly designed or malfunctions, and its output signal ends up having a wider bandwidth? You might see something like the scenario in Figure 8-6(b), where the bandwidth exceeds the channel width. In this case, the signal could start to interfere with the neighboring channels.

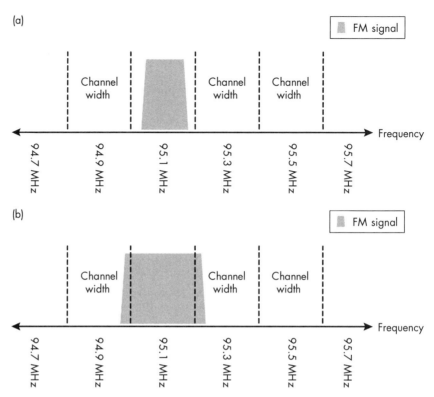

Figure 8-6: A signal positioned correctly within a channel (a) versus a signal exceeding its channel width (b)

Is this second scenario impossible? No, in fact, it actually happens on rare occasions. Is it illegal? Well, the Federal Communications Commission (FCC), or your country's regulatory body, sure doesn't like it.

Correctly designed radio systems operate such that the bandwidth of their transmitted signal is contained within an expected range of frequencies. What happens when the channel width is smaller than the signal that needs to be transmitted? This isn't a theoretical question but rather one that applies to every AM radio transmission. To see why, assume that the amount of bandwidth a modulated signal consumes is at least as large as the bandwidth of the signal going into the modulator on the transmit side. Assume also that the modulated bandwidth is proportional to that of the input signal. (There's more going on with this relationship, which we'll explore in Chapter 10, but assume this much for now.) Then recall that audio signals range between 20 Hz and 20 kHz, resulting in a 19.98 kHz bandwidth of a theoretically perfect audio signal. As you saw in your AM radio project from Chapter 4, however, the channel spacing between AM stations is only 10 kHz.

How do AM broadcasters fit 19.98 kHz of signal into 10 kHz channels? Let's find out by taking a closer look at the bandwidth of an audio signal. Create a new flowgraph and save it as *bandwidth.grc*. Add a Wav File Source and link its File property to *ch_05/HumanEvents_s32k.wav*. Then add a Low Pass Filter, setting its FIR Type to Float->Float (Decimating), its Cutoff Freq to cutoff, and its Transition Width to cutoff/10. Next, add an Audio Sink and a QT GUI Frequency Sink, with the latter's Type set to Float and its Spectrum Width to Half. To give us runtime control over the filter, add a QT GUI Range with an ID of cutoff, a Default Value of 16e3, a Start value of 500, a Stop value of 16e3, and a Step value of 500. Finally, connect the Wav File Source to the Low Pass Filter input and the Low Pass Filter output to both the QT GUI Frequency Sink and the Audio Sink. When you're done, the flowgraph should look like Figure 8-7.

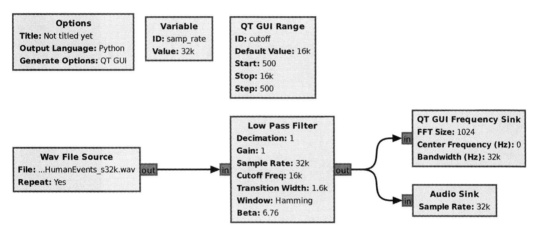

Figure 8-7: An audio filtering flowgraph

By default this flowgraph will stream in an audio clip, filter out the frequencies greater than 16 kHz, and display the resulting fast Fourier transform while playing the filtered audio on your sound card. While the flowgraph runs, we'll also have control over the filter cutoff. Run the flowgraph, which should produce some familiar sounds as well as the FFT shown in Figure 8-8.

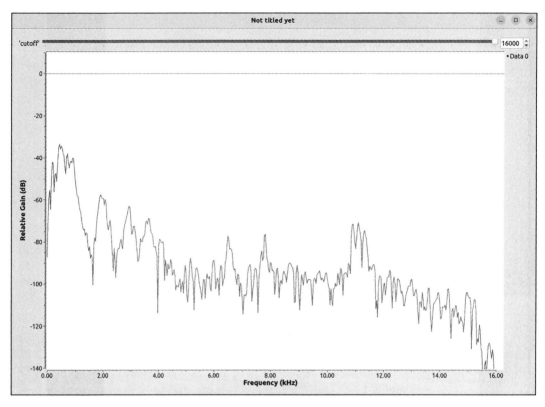

Figure 8-8: An audio FFT with a 16 kHz low-pass filter cutoff

The initial 16 kHz cutoff is close to the 20 kHz maximum frequency for audible sound, and thus the filter passes nearly all of the audio spectrum by default. (As it turns out, this file contains only frequency components up to 16 kHz, but the explanation for that will have to wait until Chapter 11.) Because the filter is not removing anything from the audio, the sound quality should be fairly normal.

Now lower the `audio_cutoff` value to make the low-pass filtering more extreme. This simulates the effect of a radio broadcaster narrowing the bandwidth of their input signal. First, try 10 kHz, which will result in a frequency plot like that of Figure 8-9.

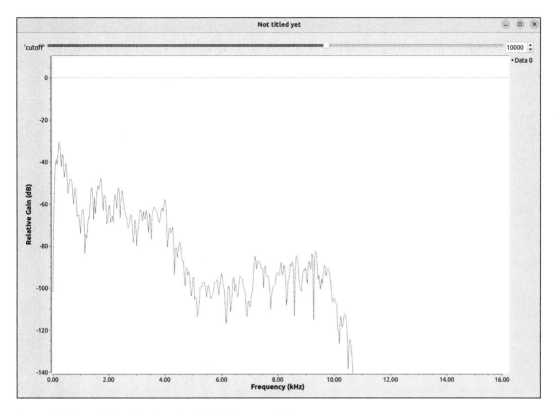

Figure 8-9: Demodulated audio after 10 kHz low-pass filtering

You should see the FFT change as you lower the cutoff value, but do you hear a difference in the audio? Maybe, maybe not. To our ears, there's a muffling effect, but it's very slight. Now try changing the cutoff to 5 kHz. You should now hear a definite muffling effect. How about 3 kHz? Even more muffling. You can still make out some of the words, even when you have only the portion of the signal at or below 1 kHz, as you see in Figure 8-10, but the narrower the bandwidth of the audio signal, the poorer the signal's quality (or *fidelity*, if we want the precise word).

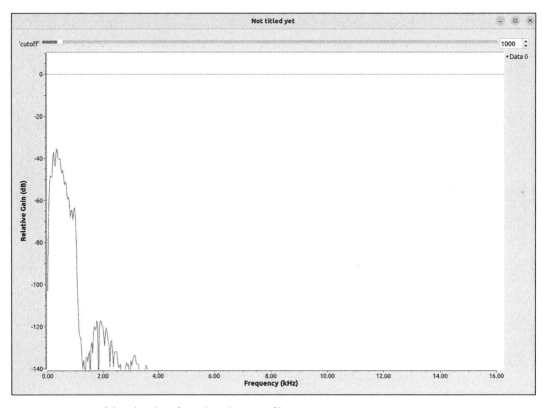

Figure 8-10: Demodulated audio after 1 kHz low-pass filtering

Based on our earlier assumption, the lower the input signal's bandwidth, the lower the modulated signal's bandwidth as well. AM radio stations take advantage of this by filtering the input signal enough to fit within the prescribed channel width, but not so much that the result sounds horrible. Filtering the input signal *before* modulating it is key to this process. Otherwise, the higher frequencies in the input signal (the ones you've decided you don't need) will still be present in the input to the modulator. Consequently, they'll cause the modulator output to consume a higher bandwidth. We'll get a closer look at this issue when we build a transmitter flowgraph in Chapter 13.

Noise

A significant portion of your time working with SDR hardware, or any kind of radio, will be devoted to fighting *noise*, by which we mean anything your receiver senses that's not the transmission you want. Noise on the RF spectrum is a lot like the noise that constantly bombards us in the audio spectrum. Close your eyes for a moment, wherever you are, and listen. No matter how quiet your room, park bench, or subway car might be, there will still be some level of audible noise. Maybe part of that noise is the hum of a refrigerator or the fans on your computer. It could also be a dog barking or cars passing by. It could even be the blood pulsing in your own ears. A perfectly quiet room doesn't exist with respect to the audio spectrum, nor does it exist with respect to the RF spectrum.

When a radio receiver tries to sense, or "hear," a transmission, other transmissions may interfere, much like the barking dog or another person talking. There could also be electrical equipment radiating electromagnetic waves. In fact, the compressor in your refrigerator is simultaneously generating both audio and electromagnetic noise, and the 60 Hz AC power going into your home or office generates RF noise that isn't so much different from the audio hum made by many appliances. Then there's noise generated by the imperfect components in your radio itself. Not so different, really, from the blood rushing through your ears. All these things combine to create an RF cacophony, like the sounds of a busy city all blurred together.

Do you know what happens when you eliminate all the audio noise around you, as when you enter an *anechoic chamber*, a room specially designed to get rid of all outside noise? At that point, you'll start hearing sounds coming from your own body, like your breathing and heartbeat. You might even find that you have a tiny bit of tinnitus you didn't know about. This is similar to the effect of internal noise in an SDR system, generated by the SDR hardware and the computer to which it's connected.

Viewing RF Noise

Let's look at a real-world example of RF noise. Create a new flowgraph and save it as *file_viewer.grc*; we'll use this to view the RF activity in our FM input file. This file of real-world RF data inevitably contains noise. Think for a moment how you would build this flowgraph with just a few blocks and give it a try without reading further.

Add the following three blocks: a File Source, a Throttle, and a QT GUI Frequency Sink. Connect them in the order listed, then link the File property of the source to *ch_07/fm_c96M_s8M.iq* and the Center Frequency (Hz) property of the sink to 96e6. Finally, set the samp_rate Variable block value to 8e6 to match the file's sample rate. When you're finished, the flowgraph should look like that shown in Figure 8-11.

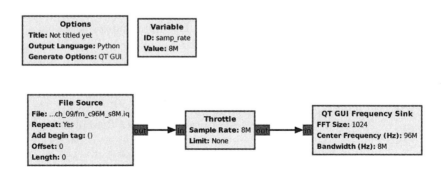

Figure 8-11: An FM receiver flowgraph with FFT attached to the source

Run the flowgraph and you'll see something like the frequency plot in Figure 8-12.

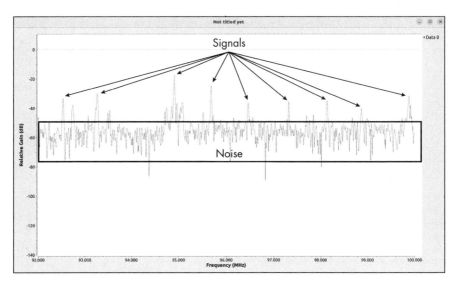

Figure 8-12: A frequency plot of the FM flowgraph RF input

The peaks in the frequency display represent FM stations, but what's going on in between? Noise, of course! This is what the RF cacophony looks like when viewed in the frequency domain. Basically, it's a minimum level of RF energy that's always present, even in the absence of a human-made signal. This level isn't still, however, but rather jumps up and down at random. We don't use the word "random" lightly. In fact, you can simulate the effects of noise in GNU Radio using a block called a Noise Source. It's essentially a random number generator that adds the random values it produces to your signal, reducing its clarity.

Finding the Signal-to-Noise Ratio

While you'll do what you can to minimize the noise level in your flow-graphs, the absolute value of the noise isn't actually critical by itself. The thing you most care about is the *signal-to-noise ratio (SNR)*, or how easily you can see your target signal relative to the noise. To go back to our audio analogy, it's not hard to listen to someone talking while in a loud, crowded room if the speaker is using a megaphone. Conversely, even in a very quiet room, it could be hard to hear someone if the speaker is whispering softly enough. It's the loudness of the voice *relative* to the noise level in the room that matters.

Looking back at the frequency plot of the FM radio input data, the signals are quite distinct relative to the noise. In fact, the strongest stations are broadcasting more than 20 dB higher than the noise level, as shown in Figure 8-13.

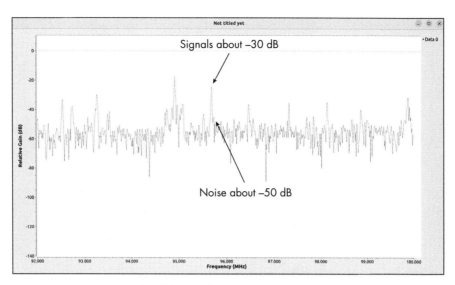

Figure 8-13: A frequency plot of the FM flowgraph RF input showing signal strength and noise level

While 20 dB may not sound like a big number, recall from Chapter 5 that decibels are a logarithmic value. A 20 dB difference therefore means the FM signal power is 100 times stronger than the noise level.

This 20 dB value is a rough estimate of the SNR for our system. It might sound odd to refer to the outcome of a subtraction operation as a ratio, but subtracting one logarithmic value from another is mathematically the same as dividing the pre-logarithmic values. Formally speaking, then, the SNR is equal to the strength of the signal (in dB) minus the strength of the noise (in dB).

Conclusion

In this chapter you learned that radio signals are carried on electromagnetic waves, the frequency of which determines how they propagate. You also saw how a signal's bandwidth represented the amount of frequency it consumes in the RF spectrum. Finally, you saw what noise looks like, as well as the importance of the SNR. With these concepts under your belt, you're finally ready to connect your GNU Radio flowgraphs to SDR hardware and work with RF signals in real time.

9

GNU RADIO FLOWGRAPHS WITH SDR HARDWARE

The wait is over: in this chapter you'll learn to interface with a real radio. You'll adapt your FM receiver flowgraph to work with actual SDR hardware so it can take in and process live radio frequency data.

Up to now, we've focused on the *software* part of software-defined radio. The inputs and outputs of your flowgraphs have largely been computer-related entities. You've extracted RF data from files and processed the data with your flowgraphs, or you've generated mathematically pure waveforms with signal sources and used them to illustrate a variety of concepts. On the output side, you've sent all of your flowgraph data to GNU Radio sinks that let you visualize your results, or you've sent the output to the sound card on your computer.

While we've been able to ease the learning process by working with known good data files and mathematical constructs, it's time to start working with the real thing: actual radio hardware. As usual, we'll dive in headfirst, then figure out what we did afterward. As you perform the conversion to hardware, keep an eye out for the parts of the flowgraph you need to change, as well as the parts that remain the same. You may be surprised to see how little the previous file-driven FM receiver flowgraph needs to change to support hardware input, but there are a few hardware-specific settings you'll need to understand.

For the purposes of this chapter, we're going to assume that you have a HackRF SDR and tailor the instructions accordingly. However, we'll also explain how the setup process differs for other SDRs, such as an Ettus USRP device or PlutoSDR, at the end of the chapter. If you have another type of SDR, you should be able to figure out which source block to use instead of the blocks discussed here and how to configure it via a quick web search.

Creating a Hardware-Enabled Flowgraph

We'll start by converting your existing FM receiver flowgraph into one that can interface with an actual HackRF SDR. First, open the *ch_09/fm_rx.grc* project. As you can see both on your screen and in Figure 9-1, it's functionally the same as the FM radio you built in Chapter 7.

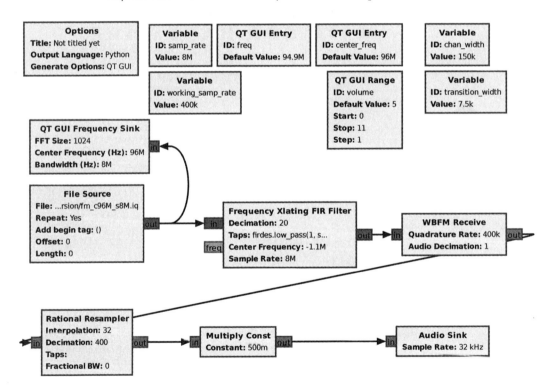

Figure 9-1: An FM receiver flowgraph with file input

You'll need to modify this flowgraph so that it gets its data from a real SDR rather than a file. Delete the existing File Source block, then add a Soapy HackRF Source, connecting its output to both the Frequency Xlating FIR Filter and QT GUI Sink inputs. As you can see in Figure 9-2, this new block is replacing the previous File Source as a supplier of radio data to the flowgraph.

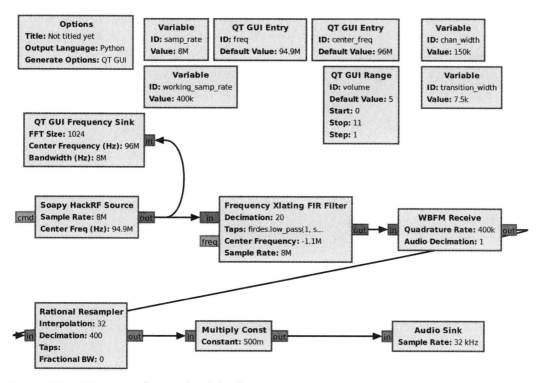

Figure 9-2: An FM receiver flowgraph with hardware input

You may notice that this new block's Center Freq is already set to a value, but it's not the 96 MHz that we need. Open the Soapy HackRF Source block's properties, click the **RF Options** tab, then change Bandwidth to samp_rate and the Center Freq (Hz) to center_freq. This tells the SDR which frequency to center on when capturing RF data. When you're done, the block's properties should look like Figure 9-3.

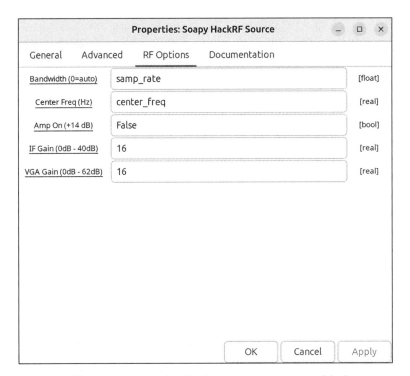

Figure 9-3: The properties window for the Soapy HackRF Source block

You're finished! All you had to do is swap your input sources, and your flowgraph will be hardware-enabled.

Setting Up the Hardware

You're ready to hook up your SDR hardware. First, attach your antenna to the HackRF board. (I'm assuming you have an ANT500, but other antennas will work.) It's best to attach the antenna to an SDR before powering it on because transmitting a signal without an antenna installed can damage your SDR. Even if you have no plans to use your SDR to transmit, this is still important, as you might end up transmitting unintentionally. For example, maybe you'll press the wrong button or select the wrong GNU Radio property. Accidents happen! To be safe, always connect the antenna first.

Where do you put it? Conveniently, there's a port on the HackRF labeled "ANTENNA" (see Figure 9-4).

Figure 9-4: A HackRF board with an ANTENNA port in the bottom-left corner

The ANT500 antenna has a connector with a roughly similar size but a different shape (Figure 9-5). These are called SubMiniature version A (SMA) connectors. The antenna's connector is SMA-male, and the HackRF's connector is SMA-female. We'll take a further look at the world of connectors in Chapter 12.

Figure 9-5: An ANT500 antenna connector

Carefully align the small pin on the antenna connector with the corresponding hole on the HackRF connector, then gently screw the outer housing of the antenna connector onto the threads of the HackRF connector, as shown in Figure 9-6. It's not so different from hooking up the coaxial connectors used by a cable TV or modem.

Figure 9-6: Attaching the ANT500 to the HackRF

Next, attach the HackRF to your computer via a USB cable. You'll need a standard male USB type A connector (the big kind) on the computer end and a micro-USB type B connector (the little kind) on the HackRF end. Once everything is connected, several of the lights on the HackRF should come on, as pictured in Figure 9-7.

Figure 9-7: The HackRF LEDs

If at any point the HackRF is unresponsive when you attempt to operate it, push the RESET button on the left side of the board. This will often allow it to recover from malfunctioning states. Stay away from the DFU button, though. Holding this down while doing other things can put your board into a mode you don't want right now. You should push this button only if you need to update your HackRF board's firmware.

Operating the Hardware SDR Receiver

Now that you've got your board plugged in and its lights turned on, you can try out the radio. Going back to GNU Radio Companion, click **Execute** to run the flowgraph. Some text similar to the following should appear in the console window to tell you about the connection that GNU Radio has made with your HackRF board (the exact details may differ):

```
Generating: "/home/paul/book/01_field_exp_sdr/ch_09/solutions/top_block.py"

Executing: /usr/bin/python3 -u

/home/paul/book/01_field_exp_sdr/ch_09/solutions/top_block.py
[INFO] Opening HackRF One #0 14d463dc2f6778e1...
```

Next, you should hear either intelligible audio or the sound of static, depending on your default tuning and the radio stations available in your area. If your default value for freq (one of the QT GUI Entry blocks) happens to coincide with a broadcast station, then you'll hear something that sounds like FM radio—either music or speech. Most likely, however, you'll need to tune the radio to a station. You should already have an idea how to do that from Chapter 6 when you practiced tuning with a raw RF input file.

Recall that the peaks you see in a frequency display represent stronger RF intensity. In the FM band, the only likely causes of these peaks are FM broadcast signals. Hover your mouse over one of the peaks and look at the frequency displayed. If you're in the United States, this frequency should be an odd multiple of 100 kHz. Enter that frequency into the freq input box near the top of the screen to hear the audio from that signal. Running this flowgraph in my local area produces the frequency plot shown in Figure 9-8, but your particular display will vary based on the FM broadcast signals present in your area.

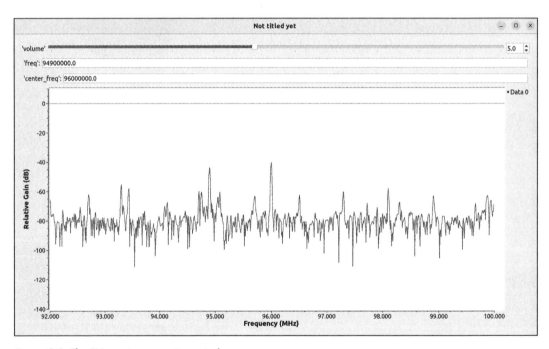

Figure 9-8: The FM receiver execution window

Take some time to poke around the FM band and see how many stations you can pick up. Keep in mind that the only difference between this hardware-driven radio and the purely software-driven models we saw earlier is the source of the input data. Here you're taking live input from the world around you via your SDR, whereas before you were taking input from a file of prerecorded RF data.

If you believe you're tuned to a radio station but you're still picking up noise or static, you may need to reposition the computer, SDR, and connecting cable.

Using USRP Hardware

The instructions outlined for the HackRF require only a few modifications to work with different SDR hardware. For example, to use an Ettus USRP, insert a UHD:USRP Source block into your flowgraph instead of a Soapy HackRF Source, and connect it in the same way you saw in Figure 9-2. Then double-click this new block to bring up its properties. Like the Soapy HackRF Source, it has several tabs along the top. Click the **RF Options** tab, then set the following:

- Ch0: Center Frequency to center_freq
- Ch0: Gain Value to 40
- Ch0: Antenna to RX2 (this is the default)

In the end, your properties window should look like Figure 9-9.

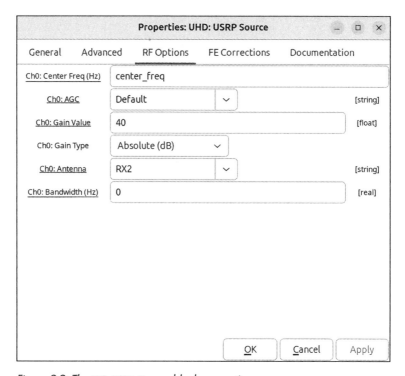

Figure 9-9: The UHD:USRP Source block properties

When completed, your flowgraph will look similar to the HackRF version, just with a different source block (see Figure 9-10).

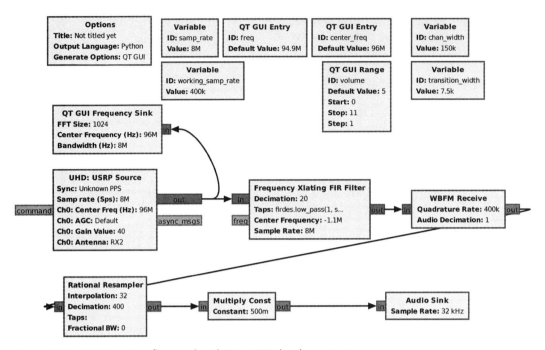

Figure 9-10: An FM receiver flowgraph with Ettus USRP hardware input

Before running this flowgraph, ensure that you have your antenna attached to the RX2 port on your USRP hardware.

Using Other Hardware

If you have another SDR, such as a PlutoSDR, bladeRF, or LimeSDR, you must first make sure that you've installed drivers for your hardware. Check with the manufacturer of your SDR hardware for installation instructions.

After that, replace the Soapy HackRF Source in your flowgraph with the source block corresponding to your hardware, connecting its output in exactly the same way. You'll then need to configure the block's properties according to your device's documentation. The properties may have slightly different names, but they will utilize the samp_rate and center_freq values. You'll typically need to supply a hardware gain value as well. Keep in mind that some platforms may not support an 8 Msps sample rate, so you may need to reduce that.

Conclusion

In this chapter you finally got to plug in some hardware! A key point we hope you realized in the process is that you can easily swap between different sources of radio data in your receiver flowgraphs. This could be switching between different files containing captured radio data or, as we've done here, switching between a file and live radio data from an SDR sink. The remainder of the flowgraph requires little to no changes.

Now that you've tried your hand at receiving radio signals with real hardware, you may be wondering about transmitting as well. For that, we'll first take a closer look at how modulation works in the next chapter.

10

MODULATION

In previous chapters, we used *demodulation* in AM and FM receiver flowgraphs to extract discernible audio signals from incoming RF data. In this chapter, we'll explore the opposite process: using *modulation* to turn a discernible signal into transmittable RF data. We won't actually transmit anything, but once you understand how modulation works, you'll be better prepared to send out signals with your SDR.

We discussed modulation at a very high level in Chapter 1, noting how it involves using a property of one signal (the information you want to communicate) to manipulate a property of another signal (the *carrier*, usually a basic sinusoid). There are three aspects of the carrier that you can change: its amplitude, frequency, and phase. We'll define each of these types of modulation in this chapter and use GNU Radio flowgraphs to illustrate how they work. Our focus will be on modulating analog signals, although we'll touch briefly on digital signals as well. Before we get to that, however, we'll start by discussing the input to the modulator: the baseband signal.

Baseband Signals

The simplest description of the *baseband* is that it's the information we actually want our radio receivers to receive. In the case of our recent AM and FM projects, for example, the baseband is some audio data. From the transmitter's perspective, you can think about the baseband signal as the waveform we're trying to send over the radio, before we do any kind of modulating.

The *band* part of the term *baseband* is a clue that we're talking about a range of frequencies. In fact, another way to think about the baseband is that it's a signal that ranges in frequency between 0 Hz and some cutoff. The cutoff frequency varies depending on the specifics of the radio system in question, meaning that the frequency range of one baseband signal will differ from that of another.

To see what this "0 Hz to a cutoff" definition looks like in practice, let's revisit our FM receiver from Chapter 7 and add a frequency plot showing the baseband. Open up the *ch_10/fm_rx.grc* file, which contains a copy of that project. Then add a QT GUI Frequency Sink and connect it to the output of the WBFM Receive block. Set the sink's Bandwidth property to working_samp_rate, its Type to Float, and its Spectrum Width to Half. Note that the Spectrum Width property won't be available until after you've set the Type. The flowgraph should now look like Figure 10-1.

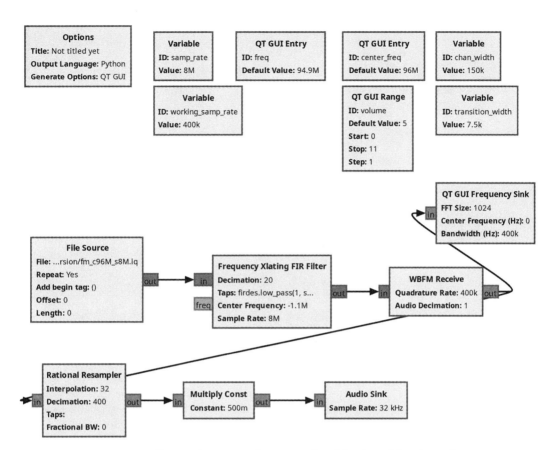

Figure 10-1: An FM receiver flowgraph to view a frequency plot of the demodulated signal

Because the frequency sink is coming out of the WBFM Receive block (the demodulator in this flowgraph), it gives us a view of the baseband: the desired signal after it's been demodulated. When you execute the flowgraph, the plot should look something like Figure 10-2.

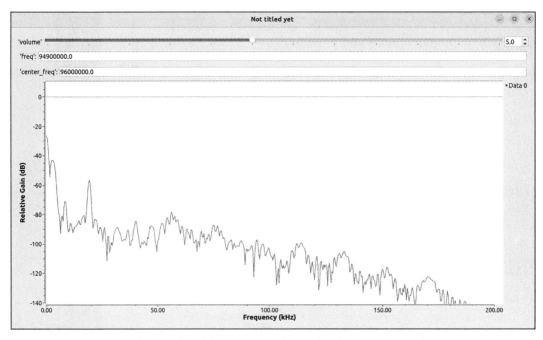

Figure 10-2: A frequency domain plot of the baseband (demodulated) signal

This frequency plot meets the requirements of a baseband signal because it ranges from 0 Hz up to a cutoff. In this case, the baseband corresponds to an audio signal (note that the signal flows into an Audio Sink after resampling and gain) containing human voices talking. Human voices almost always have most of their signal energy concentrated in the lower portion of the audio spectrum, and indeed we can see in the plot that the highest levels are near 0 Hz. Strictly speaking, the frequency plot shows the signal extending all the way up to almost 200 kHz, well above the high end of the audio spectrum (20 kHz). This part of the signal is simply noise and can safely be ignored.

If our radio was receiving digital rather than analog data, the baseband waveform would look much different and could contain much-higher-frequency components than the voice data shown in Figure 10-2. Regardless of the underlying information contained in the baseband signal, however, the key is that the baseband signal will be both the input to the modulator on the transmitter and the output from the demodulator on the receiver.

NOTE *In forms of telecommunications where many digital communication signals (voice, data, video, and so on) can occur simultaneously in a channel, the term baseband usually describes the information contained in a single channel. This definition is similar in some ways to how we've been using the term, but there are some important differences.*

Now that we know something about the baseband signals going into modulators, we can turn our attention to the modulators themselves.

Amplitude Modulation

Amplitude modulation entails scaling the amplitude of the carrier signal down when the baseband signal is low and up when the baseband signal is high. For example, Figure 10-3 shows a slower-moving baseband signal (Mod Input) overlaid on top of a faster-moving amplitude-modulated carrier signal (Mod Output). Notice how the AM signal's strength rises and falls according to the shape of the baseband.

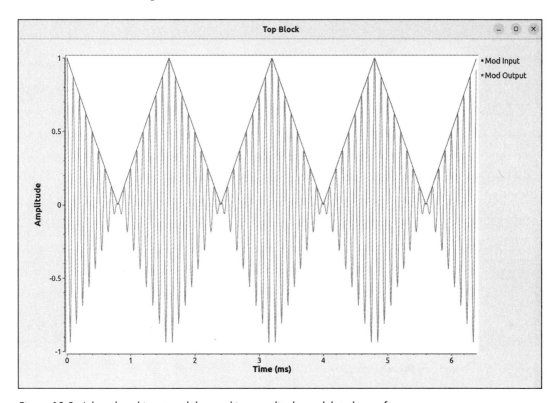

Figure 10-3: A baseband input and the resulting amplitude-modulated waveform

You can think of amplitude modulation as applying variable gain to your carrier: smaller modulator input, smaller gain; larger input, larger gain. Mathematically, this is represented as follows:

$$m(t) = bb_{in}(t) \times c(t)$$

Here m is the modulator output, bb_{in} is the baseband input to the modulator, and c is the carrier. All three of these are functions of time t, meaning their values change as the microseconds tick by. Don't worry if the math or the function notation is foreign to you; the takeaway here is that the modulator output at any point in time could be produced by simply multiplying the baseband signal level by the carrier.

This would be pretty easy to build in GNU Radio, but unfortunately, there's a problem. We'll use a flowgraph to identify this problem and then discuss how to fix it. Open the flowgraph found in *ch_10/amp_mod_begin .grc*, which contains a very simple simulated radio system, consisting of a modulator and a demodulator. The modulator input is the baseband signal, the information we're trying to send. The output of the demodulator, the receiver output in this simple system, should be the same as the baseband signal we start with. The entire flowgraph is shown in Figure 10-4.

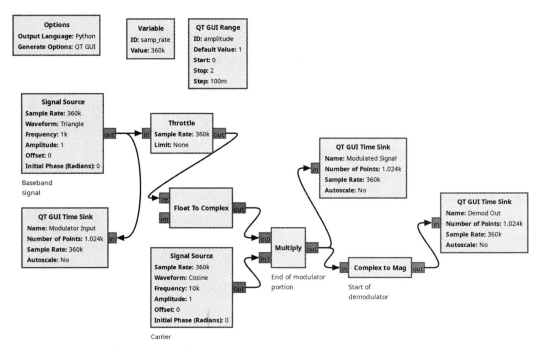

Figure 10-4: A flawed amplitude modulator

The baseband waveform in this example is a triangle wave, a waveform that transitions linearly from its minimum value (0 for the moment) to its maximum value (1) and back again. It looks like a ramp pattern that repeatedly ascends and descends. It's a much simpler waveform than would typically be used as a modulator input, but it will be useful for observing the behavior of the amplitude modulation process. The waveform, generated by a Signal Source block, is a floating-point signal, as we would expect a "real-world" baseband signal to be.

For now, the amplitude modulator consists of a Float to Complex block, which converts the floating-point samples exiting the Signal Source to complex samples (see the "Float to Complex" box for more on this block). Then the baseband ramp signal is multiplied by the cosine output of a second Signal Source, the carrier. When this cosine wave is multiplied by smaller values, it shrinks; when it's multiplied by larger values, it grows.

FLOAT TO COMPLEX

The Float to Complex block performs type conversion, converting each sample from one data type to another. This is similar to "casting" in traditional programming languages like C++. The Float to Complex block has two floating-point inputs and a single complex output. Each pair of input samples is combined into a single complex sample on the output. In mathematical terms, the first floating-point input is called the *real* part and the second input is called the *imaginary* part of the resulting complex number. Because the second input is left unconnected in the flowgraph in Figure 10-4, the block assumes 0 values for the imaginary part. The complex-typed output is therefore mathematically equivalent to the single real floating-point input.

The modulated signal leaves the Multiply block and enters the Complex to Mag block, which performs the AM demodulation. The result should be the recovered baseband signal. Unlike the AM Demod block we used in our earlier AM receivers, which also has built-in filtering, the Complex to Mag block performs just the demodulation and nothing else.

Execute the flowgraph and you'll see an output like that in Figure 10-5.

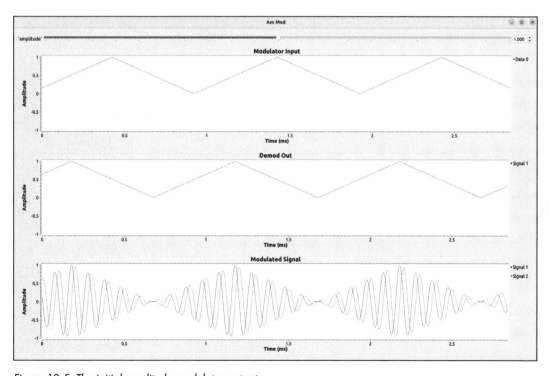

Figure 10-5: The initial amplitude modulator output

This looks pretty good, right? The modulated carrier shrinks and grows along with the size of the input baseband signal. And the demodulator output looks the same as the baseband input. There's a problem, however. The simple multiplication works because our baseband signal is taking on only positive values, since the Signal Source block's triangle pattern, by default, goes from 0 to 1 and back again. In a typical system, however, the baseband signal will take on both positive and negative values, and this complicates the process.

Working with Negative Baseband Values

When the baseband signal includes negative values, it won't be demodulated properly. To expose the issue, open up the baseband Signal Source block and change its Amplitude property to 2 * amplitude. (Here amplitude is a QT GUI Range block set to 1 by default.) Also change the Offset property to -1 * amplitude. This will produce a waveform that has the same triangular shape as before but transitions from –1 to +1. The resulting flowgraph can be seen in Figure 10-6.

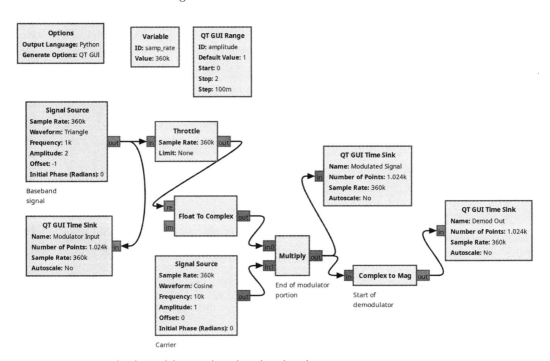

Figure 10-6: An amplitude modulator with realistic baseband input

Execute the flowgraph, and the output should look like Figure 10-7.

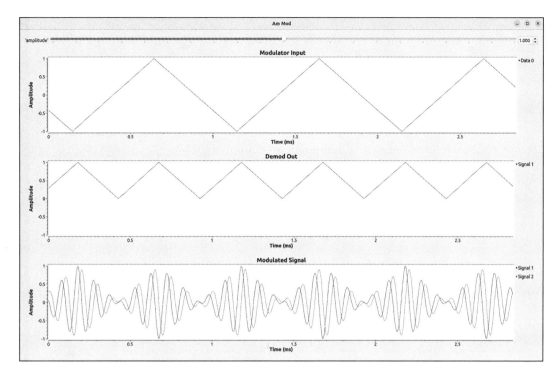

Figure 10-7: Erroneous amplitude modulator output

At first glance, you might just think the demodulated waveform is oscillating twice as fast as the baseband signal. But on closer inspection, you can see that the demodulated signal isn't going below 0. The negative portions of the original signal aren't being recovered by the demodulation process, but instead are being represented as if they're positive. You're witnessing *phase reversal*. To understand what's going on here, we're going to have to look more closely at sinusoids.

Compare the two overlapping waveforms shown in Figure 10-8. One is a sinusoid, and the other is that same sinusoid multiplied by –1.

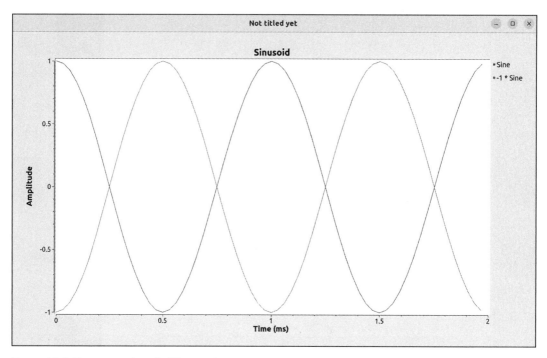

Figure 10-8: Two sinusoids with different phases

Visually, note that the difference between the two is that they're vertically flipped. Another way to describe the difference is that the two are horizontally shifted with respect to each other. There's a term for that horizontal shift: *phase*. Without getting too mathematical about it, the phase of a sinusoid has to do with its position on the horizontal axis. Shift the waveform to the left or to the right, and you've changed its phase.

One of the sinusoids in Figure 10-8 has been shifted horizontally by an amount equal to half of the waveform's total period. When the phase shifts by this amount, the waveform's phase is said to be *reversed*. Another way to reverse the phase is to multiply the original sinusoid by a negative value, which also produces the vertical flip seen in the figure.

Returning to our flowgraph, the demodulator doesn't have any awareness of the phase of the signal passed into it. It produces a result based solely on the size of the sinusoid applied to its input. As such, it sees both sinusoids in Figure 10-8 as the same, producing an identical output for both. This means that the current amplitude modulation scheme has no way of distinguishing between positive and negative baseband values.

A properly configured amplitude modulator avoids applying negative inputs to the multiplier by first shifting the baseband waveform upward on the vertical axis. If the baseband signal has a minimum value of –1, for example, we would simply add 1 to it before multiplying, so its new minimum value is 0. This would result in a new amplitude modulation equation of:

$$m(t) = (1 + bb_{in}(t)) \times c(t)$$

To add this feature to the flowgraph, break the connection between the Throttle and Float to Complex blocks and insert an Add Const block between them. Give the new block an IO Type of Float and a Constant value of 1. Then, to accommodate the larger signal passing through the flowgraph, change the Y Min and Y Max properties of the QT GUI Time Sink labeled Demod Out to 0 and 2, respectively. Also change the Modulated Signal QT GUI Time Sink block's Y Min and Y Max properties to -2 and 2. When complete, the flowgraph should look like Figure 10-9.

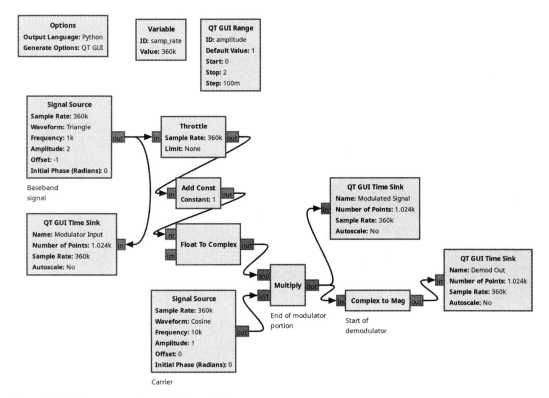

Figure 10-9: The corrected amplitude modulator

Run the flowgraph, and you should see something like Figure 10-10.

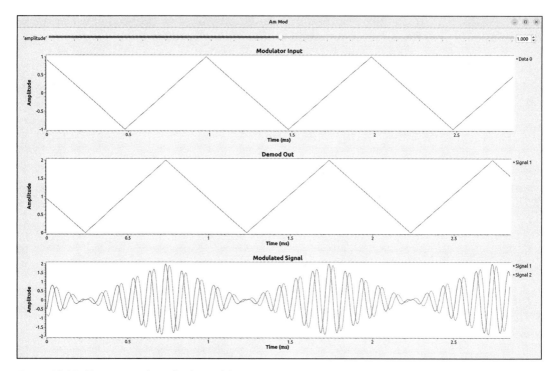

Figure 10-10: The corrected amplitude modulator output

The demodulator output now resembles the baseband input, transitioning between 0 and 2. But we're not out of the woods yet: the problem can still reemerge if the baseband signal gets too strong.

Avoiding Overmodulation

Adding 1 to the baseband signal prevents phase reversal when the signal ranges from –1 to 1, but what if the signal dips below –1? The phase reversal will come back. When the input to a modulator falls outside its legal range (in our current case, it goes below –1), it creates a situation called *overmodulation*. The result of the overmodulation is phase reversal, which introduces errors when demodulation occurs. To see this in action, use the QT GUI Range slider to change the amplitude value from 1 to 1.2 while the flowgraph is executing. The result is a baseband input that goes from –1.2 to +1.2, which produces the output seen in Figure 10-11.

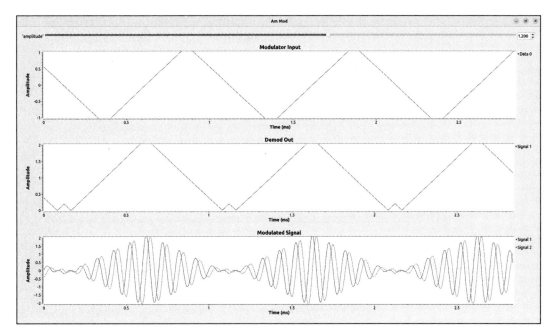

Figure 10-11: The amplitude modulator output with phase reversal

Notice how the phase reversal begins when the baseband input goes below −1. This is when the overmodulation produces the phase reversal effect. The key to avoiding overmodulation is to ensure the modulator input doesn't drop below −1, causing negative values to enter the multiplier (remember that we're adding 1 before multiplying). If you need to work with a baseband signal that has values less than −1, apply attenuation until this is no longer the case before trying to modulate it. For example, consider the excessively large signal shown in the top half of Figure 10-12 that transitions between −5 and +5.

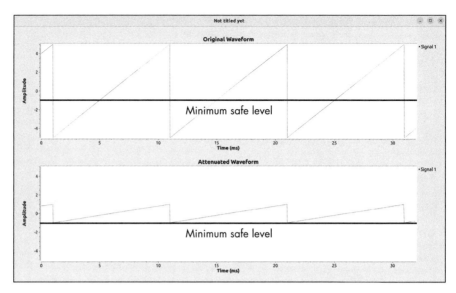

Figure 10-12: Attenuating a signal to keep it above the minimum safe level

The signal falls well below the safe level of –1. To fix that, we could apply a 5 times (5x) attenuation using a `Multiply Const` block with a Constant value of `0.2`. This would produce the signal shown on the bottom of Figure 10-12, which no longer dips below –1. The attenuated signal is now safe to pass along to the modulator.

Frequency Modulation

Frequency modulation entails altering the frequency of the carrier signal based on the value of the baseband signal. In this scheme, the frequency of the carrier speeds up when the input signal increases in size, and it slows down when the input signal decreases in size. The FM radios we implemented in Chapters 7 and 9 used this technique.

To explore how frequency modulation works, we'll use another premade flowgraph. Open *ch_10/fm_mod.grc*, which will bring up the flowgraph seen in Figure 10-13.

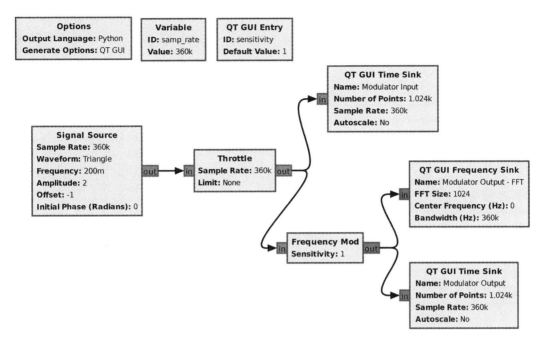

Figure 10-13: The frequency modulator flowgraph

This flowgraph contains only a modulator, rather than a modulator and a demodulator, since we won't be breaking anything like we did with the AM flowgraph. The important thing is to visualize the effect of frequency modulation on the carrier signal. In a more realistic scenario, this flowgraph might function as a transmitter, with the modulated signal passing to some kind of SDR sink for transmission to the world through your SDR hardware.

The baseband signal is a triangle wave from a Signal Source, this one transitioning between −1 and +1. The frequency of this wave is very slow: 0.2 Hz, or 5 seconds per oscillation. While a baseband signal moving this slowly is unusual, it will make it easy to visually track the behavior of the modulator. You can think of this block as steadily twisting a knob all the way from the minimum to the maximum setting and then all the way back down again.

The baseband input feeds into a Frequency Mod block, which performs the modulation. Notice that there isn't a separate Signal Source block explicitly creating a carrier signal at a particular frequency, like we had in the AM flowgraph. This isn't a mistake: the Frequency Mod block creates its own carrier for modulation, with a frequency of 0 Hz. A zero-frequency carrier may sound odd, but in fact, it's perfectly normal in an SDR flowgraph. It's a feature that distinguishes SDRs from most hardware-based systems.

Using a Zero-Frequency Carrier

In the most basic hardware radio transmitters, the baseband signal (the thing you're trying to send) will directly modulate the carrier. For example, a simple 433 MHz transmitter using frequency modulation would increase the frequency of a 433 MHz carrier as the baseband signal grew larger and decrease the carrier's frequency as the baseband signal grew smaller. Software-defined radio systems, however, don't perform modulation at the real-world carrier frequency, but instead compute the results of the modulation operation in an area of spectrum mathematically centered around 0 Hz.

Once all the flowgraph operations are completed, these zero-centered samples are sent to an SDR sink for physical transmission. The SDR hardware then uses the zero-centered data to generate a transmitted waveform at the intended carrier frequency. The center frequency property of your hardware sink (Soapy, USRP, or similar, as you saw in Chapter 9) then determines the physical frequency your radio will actually use. For example, if your transmitter flowgraph generates a signal with a frequency of +1 MHz and sends this to an SDR sink with a center frequency of 433 MHz, the flowgraph signal would be physically transmitted at 434 MHz (433 + 1). Likewise, if the output signal is at −2 MHz in the flowgraph, it would appear in the real world at 431 MHz (433 − 2).

This modulation and transmission logic mirrors what we've done with receivers and demodulators in previous chapters: the real-world frequencies captured by your SDR are determined by your SDR's center frequency property and will be denominated in MHz or GHz. The flowgraph itself, however, uses zero-centered frequencies. We'll discuss this further in Chapter 13 when we cover hardware transmission in detail. For now, let's return to the flowgraph and watch the frequency modulation process in action.

NOTE *The AM modulator in the previous section could have operated on a zero-frequency carrier as well. We chose not to build it that way since it's harder to see how amplitude modulation works without a visible carrier.*

Execute the flowgraph, and you'll see three different plots, as shown in Figure 10-14.

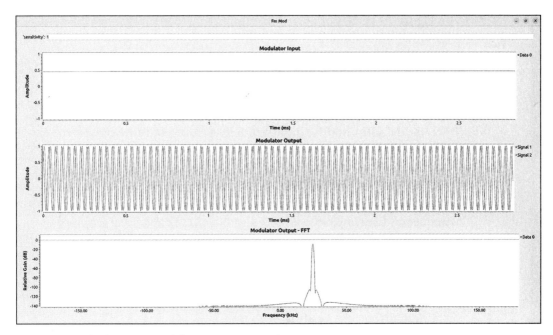

Figure 10-14: The frequency modulator execution window

The top plot shows the baseband input, which is slowly moving up and down (so slowly that it just looks like a horizontal line in the screenshot). If you count it out, you can see it move through one complete cycle in about 5 seconds. Feel free to use a stopwatch if you want to be sure.

The middle plot shows the waveform of the modulator output. The waveform's frequency is constantly changing, with the sinusoid oscillating faster and slower. It moves a bit like an accordion, squeezing together at higher frequencies and stretching out at lower ones. This movement tracks the change in the input level: when the input reaches its maximum value (+1), the modulator outputs the highest frequency; as the input passes through 0, the modulator outputs the lowest frequency; and when the input reaches its minimum value (–1), the modulator outputs what appears to be the highest frequency again.

This behavior may seem counterintuitive; why doesn't the frequency of the modulated output reach its slowest at the lowest input level? The Frequency Mod block produces an output sinusoid with a frequency proportional to the input level. If you give it a relatively high value, it outputs a relatively high-frequency sinusoid. If you give it an input value of 0, it outputs a waveform with a frequency of 0 (one that doesn't oscillate at all). Finally, when provided an input value less than 0, the Frequency Mod block outputs a sinusoid with a negative frequency.

At first glance, the time domain representation of negative frequencies doesn't look much different from the representation of positive frequencies. They are mathematically distinct, however, and this distinction can be seen on the frequency plot of the modulator output, the bottom plot in Figure 10-14. In this plot, the frequency spike perfectly tracks the input level, moving right and left as the input moves up and down and passing through zero in the middle.

This illustrates a key difference between AM and FM. An amplitude-modulated signal always appears at the chosen carrier frequency, though its size varies. A frequency-modulated signal will appear at the carrier frequency only if the modulator input is 0. In all other cases, it will be either greater than (for positive inputs) or less than (for negative inputs) the carrier frequency.

Interpreting Waterfall Plots

We've been viewing the results of frequency modulation using time- and frequency-domain plots, but there's another kind of plot that encompasses both time and frequency, providing a revealing view of what's going on during modulation: the *waterfall plot*. This type of plot shows frequency along the x-axis and time along the y-axis, creating a scrolling representation of the fast Fourier transform of a signal, like a seismograph for radio signals. The plot uses color to indicate the strength of the signal at a given frequency.

Perhaps you've already noticed the QT GUI Waterfall Sink block listed alongside the other QT GUI sinks. We'll try that block out now. Open *ch_10/fm_waterfall.grc*, a flowgraph containing the FM modulator as well as time, frequency, and waterfall sinks (Figure 10-15).

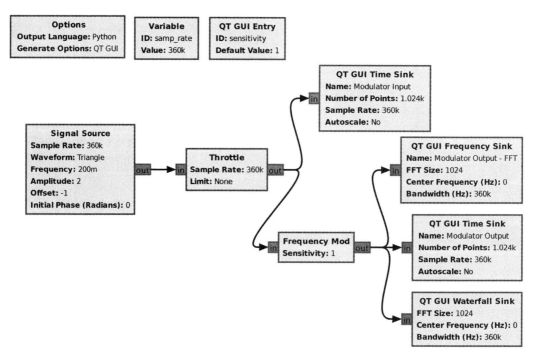

Figure 10-15: A frequency modulator flowgraph with a waterfall sink

When you run the flowgraph, you should see an execution window like Figure 10-16. The waterfall plot is the top plot in the window.

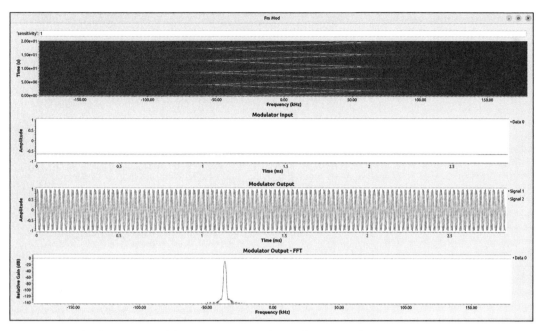

Figure 10-16: The frequency modulator execution window with a waterfall plot

Observe how the waterfall plot display scrolls upward, essentially adding a time axis to the ordinary frequency plot shown in the bottom of the execution window. The colors in the waterfall plot correspond to the vertical axis of the frequency plot, with the brighter, redder colors signifying greater FFT values and darker, more blue colors representing lower values. The frequency peak moves back and forth as the waterfall plot scrolls, tracing out a triangular pattern. You can see your actual baseband signal in the waterfall.

Waterfall plots are particularly useful when you're trying to detect short transmission pulses that you might otherwise miss on your frequency- or time-domain plots. You don't have to catch the lightning-fast blip on the FFT; you can just see a couple dots on the waterfall as they drift on by.

Adjusting Modulator Sensitivity

We've established that during frequency modulation, the modulator's output frequency changes based on the level of the baseband input, but we haven't established *how much* the frequency changes by. This is because the degree of change depends on the *sensitivity* of the modulator. This parameter of the Frequency Mod block determines how much frequency change will occur for a given input change.

To illustrate the effect of the modulator's sensitivity, the *ch_10/fm_waterfall.grc* flowgraph has a QT GUI Entry block controlling this parameter. Execute the flowgraph again, then use the textbox at the top of the execution window to increase the sensitivity from 1 to 2. Figure 10-17 shows the result.

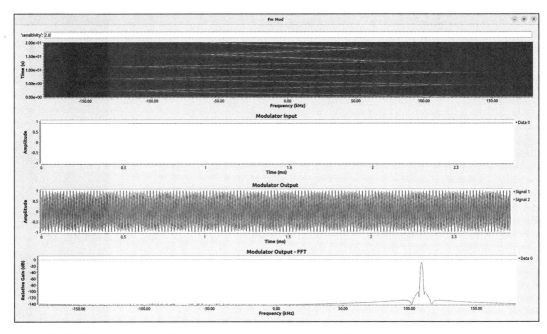

Figure 10-17: The frequency modulator execution window with a sensitivity of 2

First, notice how this change affects the frequency plot. With the higher sensitivity, the spike now moves over a wider stretch of frequencies. Meanwhile, in the time domain, the modulator output moves so fast that it's difficult to make out the individual waveform cycles. On the waterfall plot, the sensitivity change shows up as a much wider triangular range.

Now try decreasing the sensitivity to 0.2. Figure 10-18 shows the result.

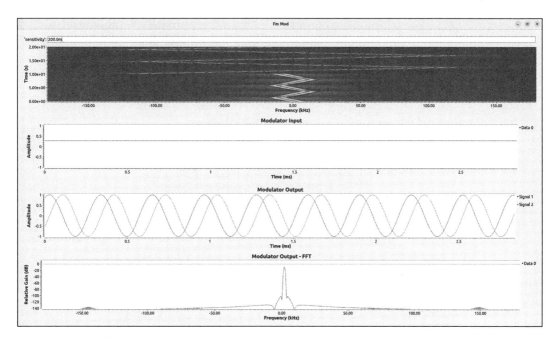

Figure 10-18: The frequency modulator execution window with a sensitivity of 0.2

With the lower sensitivity, there's a corresponding reduction in the spike's range of movement in the frequency plot and the range of the triangular pattern in the waterfall plot. In the time domain, the modulator output moves so slowly that we're no longer seeing an entire cycle of the waveform.

Another way to understand the effect of the sensitivity parameter is to look at an associated characteristic of frequency modulation: the *deviation*. This measures the maximum change in frequency relative to the carrier frequency. For example, if you're looking at an FM signal with a carrier frequency of 101.5 MHz and you observe it ranging from 101.4 MHz to 101.6 MHz, the deviation of the signal would be 0.1 MHz, or 100 kHz.

The relationship between the Frequency Mod block's Sensitivity property and the deviation is dependent on the sample rate and the maximum size of the input signal. It's often easier to scale your input signals so that they range from −1 to +1. In this simplified case, the sensitivity can be computed from the sample rate and the desired deviation (represented with Δ, the Greek letter *delta*):

$$\text{sensitivity} = \frac{2\pi \times \Delta}{\text{sample rate}}$$

For example, if your flowgraph had a sample rate of 1 Msps and you desired a deviation of 10 kHz, your sensitivity would be:

$$\text{sensitivity} = \frac{2\pi \times \Delta}{\text{sample rate}} = \frac{2\pi \times (10 \times 10^3)}{(1 \times 10^6)} = 0.0628$$

Rather than hardcode this value into your flowgraph, however, it's a better idea to create a variable containing the desired deviation (call it **deviation**) and then enter the following for the Frequency Mod block's Sensitivity property:

```
(2 * 3.1415 * deviation) / samp_rate
```

This approach allows you to change the sample rate or deviation of your flowgraph without having to manually recompute the sensitivity.

When choosing a deviation, you should mainly consider the trade-off between two factors: bandwidth consumption and ease of tuning. The bandwidth of your signal is a minimum of two deviations wide: one deviation above the carrier frequency and one deviation below the carrier frequency. (It's actually a bit wider than that due to the bandwidth of the baseband signal being modulated, but we're not going to get into the mathematical weeds on this right now). A large deviation therefore results in a large bandwidth, which means a greater risk of interfering with, or being interfered by, other signals.

On the other hand, if your deviation is very small, tuning becomes more of a challenge. Your receiver must be able to tune to a frequency that falls within the transmitted signal's bandwidth. A very small bandwidth means a very small window into which you must "aim" your tuner. This is especially important if your transmitter or receiver uses relatively cheap hardware. Such hardware typically has poor frequency accuracy, which results in tuning errors. Additionally, a larger deviation makes it easier for the receiver to demodulate the signal in a noisy environment.

Phase Modulation

The last basic modulation type is *phase modulation (PM)*. As discussed earlier in the chapter, the phase of a waveform refers to its position along the x-axis in the time domain. Therefore, with PM, we don't use the baseband signal to make the carrier bigger and smaller (like in AM), nor do we use it to make our carrier oscillate faster or slower (like in FM). Instead, we shift the carrier forward or backward in time as the baseband signal gets bigger or smaller.

We'll use another premade flowgraph to demonstrate PM. Open *ch_10/ phase_mod.grc* to view a flowgraph with a baseband input and a phase modulator, as shown in Figure 10-19.

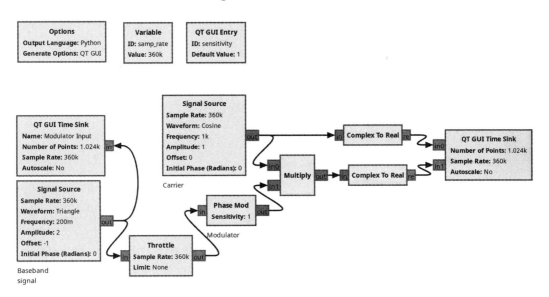

Figure 10-19: The phase modulator flowgraph

The flowgraph begins with the same slow triangle wave Signal Source for the baseband input, oscillating between −1 and +1. This input feeds a Phase Mod block, which generates a complex-valued signal corresponding to the modulator output we want. It's hard to see this phase-modulated signal, however, because its frequency is 0. To help visualize it, we've multiplied it by a 1 kHz sinusoidal carrier (the second Signal Source) to boost its frequency by 1,000. Then both the 1 kHz sinusoid and the Multiply output are fed into a QT GUI Time Sink so we can see them together in the same plot.

We haven't seen a GUI sink with multiple inputs before. If you ever want to add this kind of sink to your flowgraph, change the Number of Inputs property to 2 (or more), and the block will sprout that many inputs.

Execute the flowgraph and you'll see something like the display in Figure 10-20.

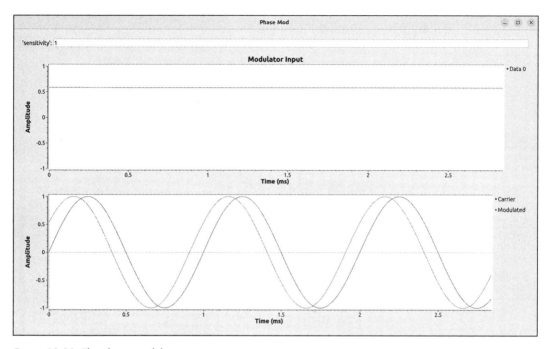

Figure 10-20: The phase modulator output

In the upper plot, you can see the familiar triangle signal moving slowly up and down. As with the FM flowgraph, it's so slow that it looks like a straight line in the figure. The lower plot shows a visual representation of PM. The stationary sinusoid represents the unmodulated carrier (the 1 kHz sinusoid before it goes through the Multiply block). The moving sinusoid is the modulated carrier. It shifts to the left when the baseband signal is positive and to the right when the baseband is negative.

When the modulated signal is to the left of the original carrier input, it's said to be *leading* the input, since a left shift on the time axis means the waveform is happening earlier in time. This is the case shown in Figure 10-20. Conversely, when the modulated signal is to the right of the original carrier, it's said to be *lagging* the input, since a rightward shift puts the waveform later in time. This case is shown in Figure 10-21.

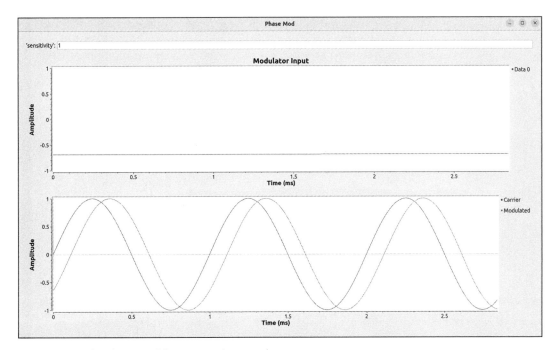

Figure 10-21: The phase modulator output lagging the carrier

For positive inputs, the modulated output leads the input by greater margins as the input gets larger. For negative inputs, the opposite occurs; the output lags the input in proportion to the magnitude of the negative input signal.

As with the Frequency Mod block, the Phase Mod block includes a Sensitivity property, which determines how much phase shift occurs for a given change in input level. Phase isn't typically used for analog modulation, however, so it's not critical to understand the mathematical specifics of how different sensitivities translate to different phase shifts.

NOTE *You may notice that the phase modulator flowgraph doesn't also contain a demodulator. This is because implementing phase demodulation using GNU Radio's recommended scheme is actually quite complicated, and thus beyond the scope of this book.*

A Word on Digital Modulation

Not all signals transmitted by a radio will be analog signals, such as voice or the gradually transitioning waveforms we've been using in this chapter. What if you wanted to send a signal containing digital information instead? Does that change the modulation process?

Think about how digital signals are generated and interpreted in the world of wired electronics. In a 3.3-volt (V) system, for example, a digital zero might be represented by 0 V and a one by 3.3 V, as shown in Figure 10-22.

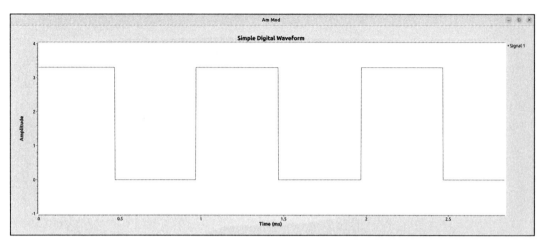

Figure 10-22: A simple digital waveform

This is a pretty simple waveform: it's either a higher value or a lower value at any given time, with nothing in between. We can apply any of the three modulation types we explored in this chapter to this waveform. For example, think for a moment what kind of modulated waveform would result if you fed this digital signal to the input of an amplitude modulator. When intending to communicate a zero, the system reduces the amplitude of the carrier to its minimum allowable size. If you scale the waveform correctly, this amplitude will be zero. Then, when the system intends to communicate the higher level, it outputs the carrier at its maximum amplitude. An example of this can be seen in Figure 10-23.

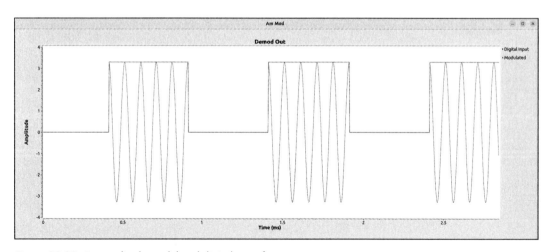

Figure 10-23: An amplitude-modulated digital waveform

You could use this system to transmit digital values, and that actually happens. This is possibly the simplest type of digital modulation scheme, and it's widely used for low-cost RF communication. It's referred to as *on-off keying*, or *OOK*. Despite having a different name, it's just a special case of amplitude modulation.

Digital modulation schemes are a topic for another book. For now, the thing to recognize is that digital modulation is built on special cases of analog modulation. If you want more of a sneak peek at what it looks like to transmit digital signals using frequency or phase modulation, go ahead and change the baseband waveform property in any of the previous flowgraphs from a triangle to a square wave, and rerun the flowgraph.

Choosing a Modulation Scheme

We've discussed three types of modulation in this chapter: amplitude, frequency, and phase. Perhaps you're wondering how to choose between them. Furthermore, why do we even need all of these different modulation schemes in the first place? Are RF engineers just inventing different techniques to keep themselves employed?

Certainly not (and never mention that conspiracy theory again). In fact, each type of modulation has its own strengths and weaknesses that make it better suited for certain applications. For an everyday illustration of how these modulation schemes differ, look no further than our old friend the car radio.

Think about the difference in audio quality between an AM and an FM radio station. You'd probably say that the FM station sounds better. But perhaps you also remember that on an interstate trip, you lose the FM stations a couple dozen miles from your home city, while the AM stations hold on quite a bit longer. They also tend to disappear in quite different ways: the FM stations quickly go from crystal clear, to cutting in and out, to completely gone. The AM stations, on the other hand, become progressively more static-laden and can slowly fade in and out at the edge of their coverage range. While the frequencies used to transmit these signals play a part in the range difference, the modulation schemes also play a part.

Most of the issue here has to do with the different frequency bands AM and FM broadcasts use; the lower frequencies of AM broadcasts propagate farther than the higher frequencies of FM broadcasts. Some of this effect is modulation related, however. Although FM is far more immune to noise when the signal is relatively strong, it's actually more susceptible to noise at low signal levels. An AM signal will get noisier as it weakens, with the noise directly showing up in the audio as the SNR drops. Because the noise doesn't directly affect the frequency of a signal, an FM signal will be very clear, almost until the FM receiver is no longer able to make sense of it, after which it will typically cut in and out and then disappear. This is one reason amateur radio operators choose variants of AM rather than FM to transmit over extremely long distances. Conversely, to transmit over relatively short distances, FM is often the better choice.

Meanwhile, different variants of PM are especially well suited for digital transmissions, which come with their own considerations. To make matters even more complicated, some digital modulation schemes use a combination of AM and PM, but that's a topic for another day.

Conclusion

This chapter has demystified the modulation process, using flowgraphs to illustrate what's happening during amplitude, frequency, and phase modulation. Ultimately, modulation is nothing more than GNU Radio performing mathematical operations on the baseband signal (the one we're trying to send). The key considerations are which modulation type to use and how to set the modulation parameters. For example, we've seen in this chapter how to attenuate a baseband signal to keep it within a range of –1 to 1 for AM and how to use the sensitivity parameter during FM to control the bandwidth of the resulting signal.

In the next chapter we'll take a closer look at how SDR hardware actually works. Coupled with your understanding of modulation, this deep dive will help you feel confident using your SDR to both receive and transmit.

11

SDR HARDWARE UNDER THE HOOD

In this chapter, we'll explore how SDR hardware like the HackRF One is able to take in radio frequency signals from the outside world and turn them into data that your computer can work with. When we first encountered SDR hardware in Chapter 9, using it in GNU Radio Companion was relatively simple: you just dropped in a block to talk to the hardware and configured a few of the block's properties. To have the most success with SDR, however, it's helpful to have a deeper understanding of how these devices work and how best to use them.

Classic Radios vs. SDR

To understand how SDR hardware works, it's useful to first examine the old-school, fixed-function style of radio design. We aren't going to spend a lot of time looking under the hood of traditional radios, but a quick peek will help clarify some important radio terms. Figure 11-1 shows a block diagram for a traditional receiver. You'll often see this kind of receiver described as *superheterodyne* (or *superhet* for short).

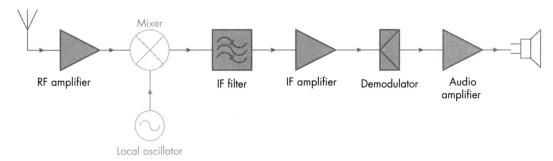

Figure 11-1: A basic superheterodyne radio

One pleasant surprise should be that none of the receiver's components are a complete mystery at this stage. You've seen amplifiers, filters, and demodulators before in this book. You've also seen mixers, though you may have forgotten that *mixer* is just another word for *multiplier*. This really isn't so different from the much simpler radio model in Figure 11-2 that we've been using to understand SDR.

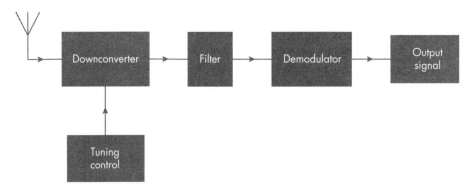

Figure 11-2: A basic radio model

Leaving out the gain (amplifier) stages for a moment, both systems have a similar structure. They both tune by downshifting the signal and then filtering it. Both systems then use a demodulator to recover the information signal, which is the original signal that the transmitter was intended to communicate. The only real differences are in some of the details, particularly that little *IF* abbreviation on the blocks in the middle of Figure 11-1.

It stands for *intermediate frequency*, and unlike our previous flowgraphs, this frequency isn't 0 Hz. Instead, superheterodyne radios use a variable local oscillator to downshift the incoming RF signal to a fixed lower frequency; no matter what station you're tuning into, the IF will always be the same. This downshifting step, known as *heterodyning*, is the defining characteristic of the superhet radio design.

Why is this intermediate frequency necessary? It's difficult to create a hardware radio where every stage has a varying range of signal frequencies passing through it. Shifting to a fixed intermediate frequency means that only the initial stages of the design have to work with variable frequencies. After that, all processing can be done with fixed frequencies. In other words, you don't have to realign all the downstream electronics every time you tune to a new signal, and this makes the radio much simpler and cheaper to design. Figure 11-3 divides the superhet design into two sections: the stages that support variable frequencies (left box) and the fixed-frequency stages (right box).

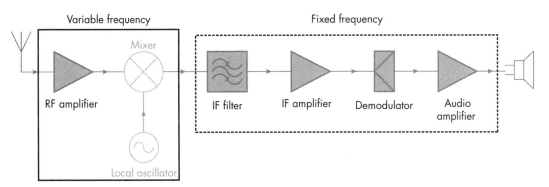

Figure 11-3: The variable- and fixed-frequency portions of a superhet radio

We don't need to spend much more time with this model. The main point is to observe how much simpler it is to implement receivers using software-defined radio. If you're trying to replicate the function of an existing radio and you read about its intermediate frequency, just know that you don't need to worry about that. Thanks to the SDR hardware, you can assume a "direct conversion" from RF frequencies down to 0 Hz, without needing to consider any kind of intermediate frequency.

NOTE *If you look at the block diagram for your specific SDR hardware, you may find that it contains a superheterodyne structure with an intermediate frequency. The HackRF One contains such a structure, for example. There are sometimes engineering reasons for using this topology in SDR hardware designs, but you don't need to worry about how the engineers designed your SDR to use it. Whatever design choices they made, your SDR looks to you, the user, like a direct conversion device.*

IQ Sampling

Arguably the main purpose of SDR hardware is to generate a stream of complex numbers that represent a portion of the RF spectrum. Once you have those numbers, you can set your computer loose on them, demodulating, decoding, and doing anything else you want to do. This means we need to consider something we've been avoiding until now: How do we get complex-valued signals from real-world phenomena?

The answer is a technique called *IQ sampling*, or sometimes *quadrature sampling*. This technique is the cornerstone of SDR. It uses two analog-to-digital converters (ADCs) to measure (sample) two different versions of the incoming waveform. Specifically, the first ADC samples the *in-phase (I)* version of the signal, while the second samples the *quadrature (Q)* version, which has a different phase. Sampling these two different phases of the same signal produces a set of complex values that provide a more useful representation of the signal than values from just one ADC could. Additionally, these complex values encompass both positive and negative frequency components.

The mathematics behind IQ sampling could fill a whole book on digital signal processing and are beyond the scope of our current discussion. Instead, this section will give you a simplified, high-level view of how the process works and provide a foundation to build on should you choose to research the topic yourself. There are several great tutorials about IQ sampling on the Web, with plenty of mathematics, if you're so inclined.

IQ Signals

Before getting to the *sampling* part of IQ sampling, let's look at the I and Q signals themselves. The key difference is that the quadrature version of a signal is 90 degrees out of phase with the in-phase version. (The *quad* in *quadrature* refers to the fact that 90 degrees is one-quarter of a full 360-degree cycle of a sinusoid.) Without getting into the mathematics, the process of multiplying a signal by a sinusoid can produce changes in phase as well as the changes in frequency we've already seen. We produce the I component by multiplying the input RF signal by a cosine wave. The cosine function is mathematically defined to have zero phase, and as such, it produces the in-phase portion. We produce the Q component by multiplying the input RF signal by a sine wave. A sine wave is always 90 degrees out of phase with its associated cosine wave, so it gives us the quadrature portion. Figure 11-4 shows the two sinusoids and their phase difference. Notice how the sine wave peaks a quarter of a cycle after the cosine wave.

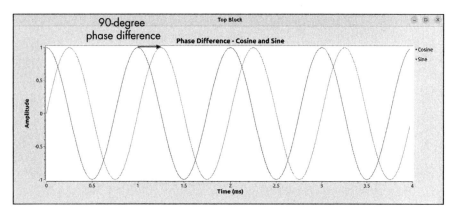

Figure 11-4: The phase relationship between cosine and sine waves

The block diagram in Figure 11-5 shows how these two different sinusoids work in parallel in an IQ sampler to produce the I and Q components of the input.

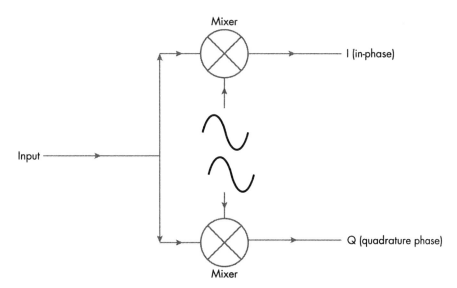

Figure 11-5: The mixing stage of an IQ sampler

The cosine and sine wave are both set to the center frequency (the frequency to which the SDR is tuned). Each sinusoid is multiplied by the input signal using a mixer, which has the same down-conversion effect as the mixer in the superheterodyne design from Figure 11-1. The math behind these hardware mixers is a bit more complicated than the software multipliers we've used in our flowgraphs, but the key point here is that the two mixers shift the input RF energy down to 0 Hz, albeit with two different phases.

Analog-to-Digital Conversion

So much for the *IQ* part of IQ sampling. For the *sampling* part, a pair of ADCs sample the I and Q waveforms to produce the complex digital data your flowgraphs will need. You might think we could do that immediately after the signals emerge from the mixers, as in Figure 11-6.

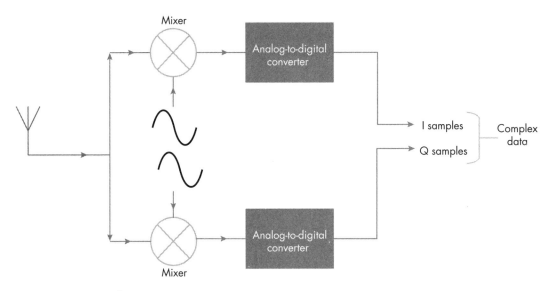

Figure 11-6: IQ sampling in concept

This almost works, but there's a problem we need to address. The ADCs can only convert their inputs accurately from analog to digital over a limited frequency range, known as the *capture window*. If signal energy with a frequency outside this capture window enters the ADC, it will distort the signal you're trying to convert. We alluded to this problem in Chapter 2 when we said that in a properly functioning ADC, the target signal must not be moving too fast to capture. Unfortunately, the I and Q inputs to the ADCs have *all* the RF energy picked up by the antenna over an infinite frequency range. Even though that energy has been downshifted by the mixers, much of it still remains at high enough frequencies to be outside the capture window, and so this extraneous RF energy will distort the analog-to-digital conversion process.

The size of an ADC's capture window is determined by the *Nyquist-Shannon sampling theorem*. This theorem is often summarized as "you need to sample at twice the frequency of the analog input signal to avoid problems." A more precise definition of the theorem states that a digital sampling system can only capture a faithful representation of an analog input signal if the input signal is band-limited to half the sampling rate. *Band-limited* simply means that the frequency components of the input signal must all fall within a finite range. In this case, the upper limit of that range is $f_s/2$, where f_s is the sampling rate. This limit is also called the *Nyquist frequency*.

Because we're talking about IQ samples, which contain information about both the positive and negative frequencies, the lower limit of the capture window is $-f_s/2$.

Figure 11-7 shows a frequency plot of a well-behaved signal that sits within the safe capture window of the ADC. The plot has no frequency components greater than $f_s/2$, nor does it have any frequency components less than $-f_s/2$, so the signal will be sampled correctly.

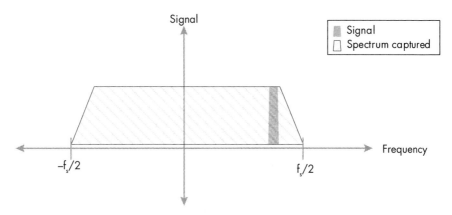

Figure 11-7: Sampling a signal within the capture window

Figure 11-8 shows another example frequency plot. This time the signal has too high a frequency and falls outside the capture window. It won't be sampled correctly.

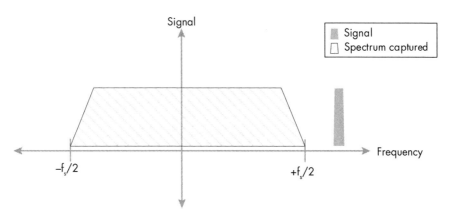

Figure 11-8: Attempting to sample a signal outside the capture window

What happens when the signal frequency is higher than the Nyquist frequency? Something called aliasing.

Aliasing

Aliasing is a phenomenon that causes frequencies higher than the Nyquist frequency ($f_s/2$) to be rendered as if they were frequencies lower than the Nyquist frequency when a signal is sampled. Once again, we alluded to this problem earlier in the book. In Chapter 5, when we were playing with basic sinusoids, we observed that as a sinusoid's frequency rises, the resulting tone eventually becomes too high to hear. But then, as the frequency continues to rise, the tone reemerges and, paradoxically, gets lower and lower. That was aliasing in action.

To see the effect in more detail, let's try a little experiment. You'll run a flowgraph that intentionally breaks the conditions demanded by the Nyquist theorem and see what happens. Open *ch_11/nyquist.grc*. This flowgraph, also shown in Figure 11-9, generates a variable-frequency sinusoid, which goes into a Keep 1 in N block, after which it is displayed in QT GUI Sink, a multifunction display block containing a tabbed interface with several different types of plots.

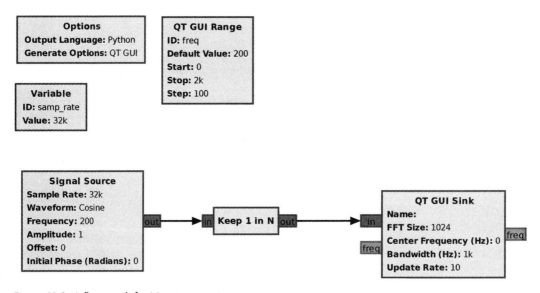

Figure 11-9: A flowgraph for Nyquist experiments

We used the Keep 1 in N block before in the decimation flowgraph from Chapter 6. It performs decimation by passing along only one out of every N samples from its input to its output. In this project, N has been set to 32, meaning that samples exit the block at a rate of 1,000 samples per second, 32 times slower than the 32,000 samples per second rate of the Signal Source. The Nyquist frequency ($f_s/2$) for any blocks downstream of the Keep 1 in N block is half of 1,000 Hz (that is, 500 Hz).

The QT GUI Range block allows you to change the frequency of the input signal, and consequently the Keep 1 in N block output, from legal values less than the Nyquist frequency (499 Hz) to illegal values greater than or equal to the Nyquist frequency (500 Hz and up). When you execute the flowgraph, you should initially see a single peak at 0.2 kHz (200 Hz), which is the default value of freq. Figure 11-10 shows this output.

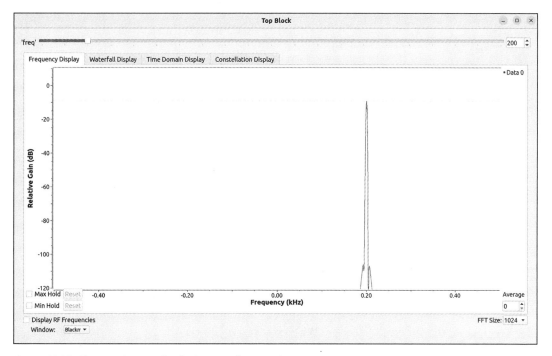

Figure 11-10: The initial output for the Nyquist flowgraph

At first, as you slide the freq control to the right, you should see the peak move to the right. Then it should straddle the right and left sides of the plot when you reach 500 Hz, as in Figure 11-11. You're now on the edge of legal territory, no longer generating valid samples.

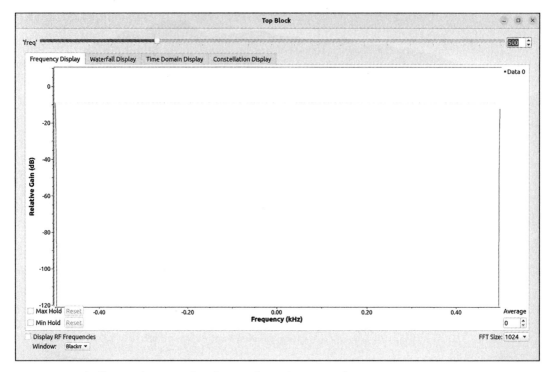

Figure 11-11: The flowgraph output when the signal is at the Nyquist frequency

Keep increasing the slider to 600 Hz. The aliased frequency should completely roll off the right edge of the window and start shifting to the right from the window's left edge. When you reach 600 Hz, the frequency plot should show the aliased signal's frequency is –400 Hz, as shown in Figure 11-12.

If you further increase the freq slider to 700 Hz, the peak will continue to move to the right, to –300 Hz. The rightward shift will continue until you get all the way to 1,500 Hz, after which another rollover will occur and the peak will jump back to the left edge of the window.

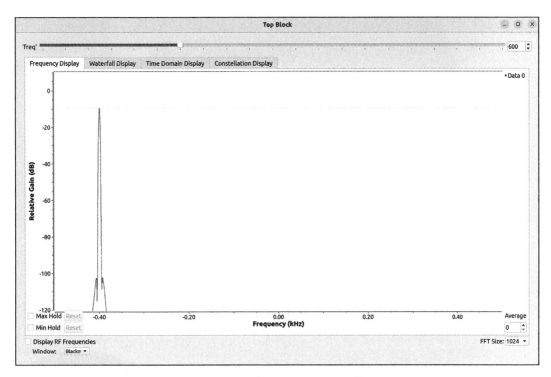

Figure 11-12: The flowgraph output when the signal is past the Nyquist frequency

In Chapter 5, we noted that the pitch of the tone started to *decrease* when aliasing started. You might therefore be surprised to see the aliased frequency continually moving from left to right as you turn up the slider. In fact, what was happening in Chapter 5 was that the frequency of the tone had "rolled over" into the negative frequencies, much like we're seeing here. For real-valued signals, positive and negative frequencies sound the same, so moving from the left edge of the window toward the right resulted in the absolute value of the tone's frequency dropping until it hit 0 Hz. From there, the pitch would rise again.

Now slowly drag the slider to the right until freq is set to 1100. Observe how the peak moves from the left side of the frequency plot, past zero, and continues right until it reaches 100 Hz.

Lest you think this strange behavior in the frequency plot is merely some quirk in the way the fast Fourier transform is computed, switch over to the Time Domain tab of the execution window. Then click and drag to zoom in on a small portion of the display, as shown in Figure 11-13.

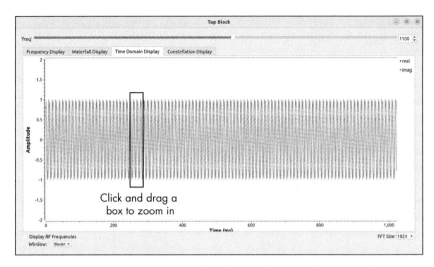

Figure 11-13: The time domain output when the signal is past the Nyquist frequency

Next, click the **imag** label to hide the imaginary part of the complex waveform, leaving only the real part. This should reveal a simple sinusoid, as shown in Figure 11-14.

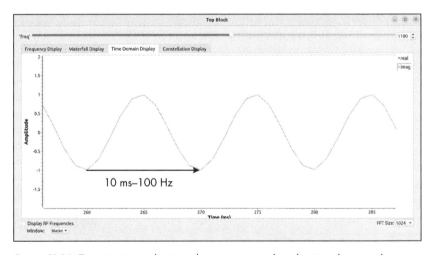

Figure 11-14: Zooming in on the time domain output when the signal is past the Nyquist frequency

The period of this waveform is about 10 ms, which corresponds to a frequency of 100 Hz, even though you're generating an 1,100 Hz signal, which should have a period of about 0.9 ms. This is the aliasing effect in action. When the 1,100 Hz signal is sampled, it gets mistaken for a lower frequency. It's as if the signal's frequency is taking on an "alias" of 100 Hz.

This experiment illustrates why you need to band-limit your signals before sampling them, making sure the ADC inputs don't contain portions that have a higher frequency than half the sample rate. Without this precaution, you'll have all sorts of aliased frequencies undergoing the kind of frequency shift you just witnessed. The shifted frequencies will overlay your real signal and cause the real signal to be distorted. The frequency domain perspective of this situation is shown in Figure 11-15.

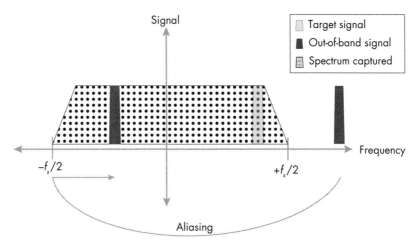

Figure 11-15: A frequency domain representation of aliasing

In the last step of our experiment, the sample rate was 1 ksps and the sinusoid being sampled had a frequency of 1,100 Hz. The signal was then significantly higher than the Nyquist frequency ($f_s/2$) of 500 Hz, and therefore aliasing occurred. One way to think of aliasing is that in a sampled system, you're only able to work with a range of frequencies equal to your sample rate (from $-f_s/2$ up to $+f_s/2$). If you ever exceed the Nyquist frequency, then the signal energy at those "too-high" frequencies will "roll over" the edge of the range and appear again on the opposite side. If you continue increasing the frequency of the input signal, it will move further to the right until hitting the $+f_s/2$ limit and rolling over again.

Try moving the slider around to get a feel for how aliasing provides an inaccurate representation of the real input signal. In general, however, you won't have to predict the results of aliasing: you just want to avoid it.

Filtering

To ensure the input signal is band-limited, the SDR hardware must use a low-pass filter with a cutoff no greater than the Nyquist frequency ($f_s/2$) to eliminate those pesky higher-than-Nyquist frequencies. In particular, the SDR hardware must use an *analog* filter because the filtering must occur *before* it performs sampling, taking digital options off the table. Properly filtering before sampling eliminates possible aliasing, as shown in Figure 11-16. As long as the signal components above the Nyquist frequency fall within the stopband of the filter, they won't alias into your signals during sampling.

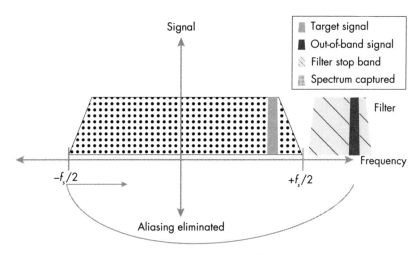

Figure 11-16: A frequency domain representation of low-pass filtering to eliminate aliasing

This filtering step was the missing piece of the IQ sampling block diagram from Figure 11-6. Adding in two low-pass filters, one for the I component and one for the Q, gives us the updated block diagram in Figure 11-17.

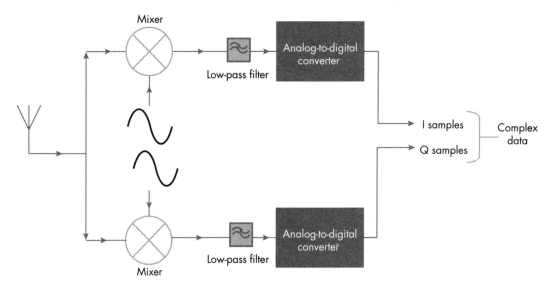

Figure 11-17: An updated IQ sampling block diagram

Typically, the sampling rate of an SDR is variable, so the low-pass filter cutoff in the IQ sampler is also variable to match. For some SDRs, you won't have to explicitly set the bandwidth in GNU Radio, as they will automatically adjust to match the sample rate. For the Soapy sources, however, you should manually configure the bandwidth to match the sample rate. We did this in Chapter 9 by setting the Bandwidth property in the RF Options tab to samp_rate.

SDR Bandwidth and Sample Rates

An SDR's bandwidth (that is, the range of frequencies the SDR can capture) is equivalent to the sample rate. You saw this when operating the FM hardware receiver flowgraph in Chapter 9. That flowgraph had a sample rate of 8 Msps, and the QT GUI Frequency Sink showed an 8 MHz range of frequencies.

The connection between bandwidth and sample rate matches the sampling theory we just discussed. We've established that an IQ sampler captures an amount of spectrum equal to the Nyquist frequency on the positive part of the horizontal axis, as well as an equivalent amount of spectrum on the negative side, as shown in Figure 11-18. Since the Nyquist frequency is equal to half the sampling rate, the total capture bandwidth (two Nyquist frequencies) is equal to the sampling rate itself.

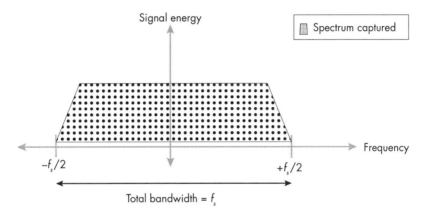

Figure 11-18: An IQ sampler frequency plot with an annotation of the total bandwidth

The relationship between sampling rate and bandwidth is perhaps the most important concept to remember from this chapter. Setting the sample rate and center frequency for your SDR establishes the capture window, or range of frequencies that will be converted to IQ data for your flowgraph. This region has a size (bandwidth) equal to the sample rate and is centered on the SDR's center frequency.

Identifying Bandwidth Limits

Three main factors limit the bandwidth of an SDR system: the speed of its ADCs, the speed of the SDR-to-PC interface, and the speed of the computer that's doing the GNU Radio processing. The first two are hard limits. The specification for your SDR will reflect both of these considerations in the maximum sampling rate of the device. For example, the HackRF specifies a maximum sampling rate of 20 Msps, a value determined by the sample rate of its ADCs as well as its USB 2.0 interface.

The third limitation, your computer speed, is more variable and harder to quantify. It depends on your CPU speed, the performance of your storage media, whether you have other programs running at the same time as GNU Radio, and the flowgraph you're running. The maximum speed of your flowgraph will depend on what kind of computational bottlenecks it has and how well equipped your computer is to handle those types of bottlenecks. It may or may not be less than the speed of your SDR.

Experiencing Overflow

When data enters your flowgraph faster than the blocks in the flowgraph can process it, the result is an *overflow* condition. To see this effect in action, open the *ch_11/fm_rx_ovf_hackrf.grc* flowgraph, which can be seen in Figure 11-19.

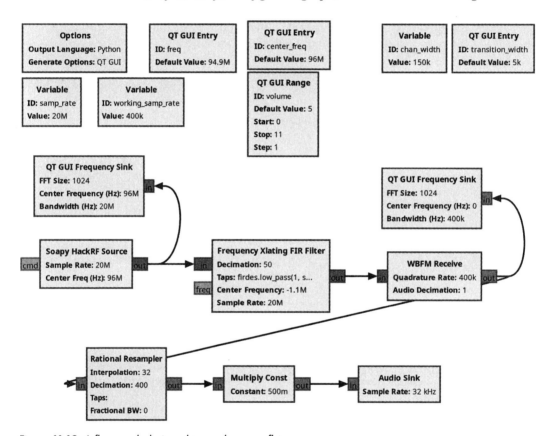

Figure 11-19: A flowgraph designed to produce overflows

This flowgraph is similar to your previous hardware-based FM radio receiver, but with the samp_rate increased to 20e6 (instead of 8e6) and a QT GUI Entry to dynamically control the transition width of your tuning filter. Also, the Default value of the transition width is smaller than the previous value. Before executing, set the Default Value for the freq variable to a frequency corresponding to a known FM station in your area. It may be hard to tune after the flowgraph runs, for reasons you'll see in a second.

Unlike the previous projects, we can't tell you what will happen when you execute this flowgraph. Even if you're running a solidly mainstream computer as of 2025, you should see and hear moderate problems. The GUI will appear, displaying 20 MHz of bandwidth, complete with the expected FM radio peaks, as shown in Figure 11-20.

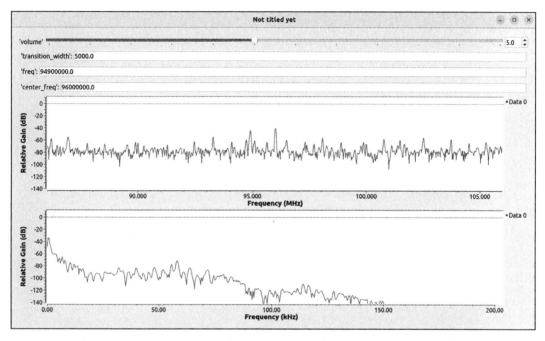

Figure 11-20: The FM receiver execution window with 20 MHz of bandwidth

Unless you have a very fast computer, however, you should hear noticeably choppy audio. You should also see some interesting output in the console pane at the bottom left of the main GNU Radio Companion window, similar to the following:

```
Generating: "/home/paul/book/01_field_exp_sdr/ch_11/fm_rx_ovf_hackrf.py"

Executing: /usr/bin/python3 -u
/home/paul/book/01_field_exp_sdr/ch_11/fm_rx_ovf_hackrf.py

[INFO] Opening HackRF One #0 14d463dc2f6778e1...
0s00s00s00s00s00s00s0aU0s00s00s00s00s00s00s00s00s00s00s00s00s00s00s00
s00s00s00s00s00s00s0aU0s00s00s00s00s00s00s00s00s00s00s00s00s00s00s00s
0os00s00s00s00s00s00s0aU0s00s00s00s00s00s00s00s00s00s00s00s00s00s00s0
0s00s00s00s00s00s00s00s0aU0s00s00s00s00s00s00s00s00s00s00s00s00s00s00s00
s00s00s00s00s00s00s00s0aU0s00s00s00s00s00s00s00s00s00s00s00s00s00s00s
0os00s00s00s00s00s00s00s0aU0s00s00s00s00s00s00s00s00s00s00s00s00s00s0
0s00s00s00s00s00s00s00s0aU0s00s00s00s00s00s00s00s00s00s00s00s00s00s00
--snip--
```

A capital 0 in the console pane indicates that your flowgraph is experiencing an overflow condition. You may also see a lowercase a or s. If a small handful of these characters appear intermittently, they may just be little hiccups that don't noticeably affect the flowgraph's function. If, however, you see a consistent stream of these characters, as shown here, you have a problem: your computer isn't processing fast enough to keep up with the incoming data.

If you don't experience significant overflow behavior, reduce the value of transition_width to 4000. As mentioned in Chapter 5, smaller filter transitions require the CPU to work harder. This change should produce overflow events, even on a very fast CPU. For those of you using *extremely* powerful machines, however, you'll need to turn up the heat even further to see an overflow condition. But how do you do that? The HackRF spec says you can't sample faster than the 20 Msps rate we're already using, so what can you do to make this flowgraph more computationally challenging?

Think about one of the rules we discussed in Chapter 2: "Never sample faster than you need to." Thanks to this rule, we've learned to decimate a flowgraph's data stream at the earliest opportunity so that blocks downstream can operate more slowly. In fact, the current flowgraph has only three blocks running at the initial 20 Msps sample rate: the Soapy HackRF Source, the Frequency Xlating FIR Filter, and a QT GUI Sink (highlighted in Figure 11-21).

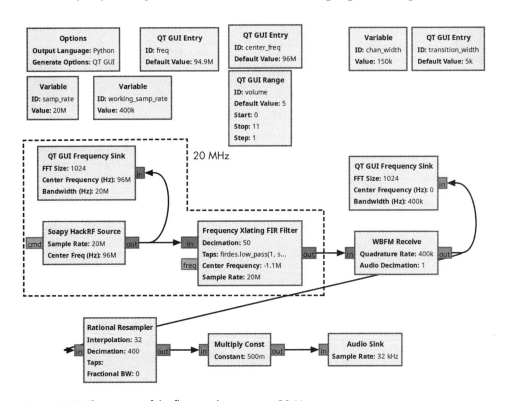

Figure 11-21: The portion of the flowgraph running at 20 Msps

Everything downstream from these three blocks runs 50 times slower, at the working_samp_rate value of 400 ksps. If you want to make this flowgraph more difficult to run, all you need to do is violate best practices and run the downstream blocks faster than necessary. For example, try changing the working_samp_rate property to 1e6. This will drive a change to the downstream sample rates, bringing some blocks up to 1 Msps (Figure 11-22).

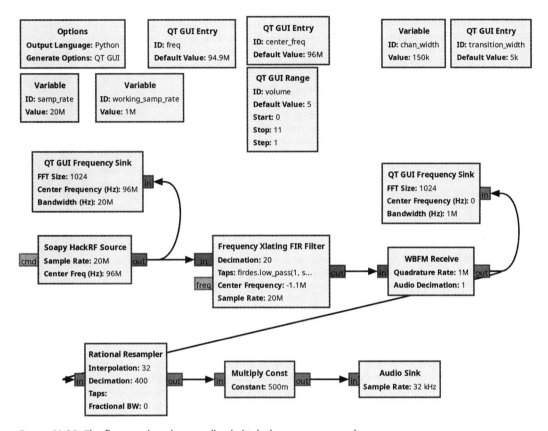

Figure 11-22: The flowgraph with a needlessly high downstream sample rate

When you run the updated flowgraph, you'll almost definitely hear completely garbled audio (or no audio) and see a very fast stream of 0 and s characters in the console, even if you're running on a very fast machine. This flowgraph is much harder for the computer to handle because all but the very last part is running 2.5 times faster than before. Hopefully this shows you how large of a difference a well-designed flowgraph can make over a carelessly constructed one.

Up to now, we've focused almost entirely on receiving, but a similar set of constraints hold for the transmit side of an SDR as well. Instead of overflowing when your computer is too slow, however, you'll have something called *underflow*. When transmitting, the SDR needs data coming from the computer at whatever sample rate you've selected for your SDR sink block. When the computer can't keep up, the SDR has nothing to transmit for a time, so it can only send "dead air." When this happens, the console pane will start displaying U characters. As with 0 characters, a small handful of these won't be fatal, but a swift progression of them means the signal you're trying to transmit has been seriously compromised and will need to be redesigned for better efficiency.

NOTE *A similar phenomenon to a transmission underflow is an* audio underrun *issue, where the audio playback hardware doesn't receive data fast enough. GNU Radio Companion will indicate this by displaying* aU *characters in the console pane.*

Preventing Overflow

If you're having overflow (or underflow) problems, you need to either reduce the computational load of your flowgraph or increase the computational power of your computer. We recommend the following troubleshooting steps, in order:

1. Optimize the sample rates throughout the flowgraph, reducing them anywhere you can by decimation.

2. Reduce the sample rate of the flowgraph's sources if possible. For example, does an FM radio really need to work with 20 MHz of spectrum at the same time? The radio still works if you sample only at 8 MHz or even slower.

3. Disable unnecessary blocks in your flowgraphs, such as GUI sinks. Sometimes when debugging a flowgraph, you'll find yourself littering it with different GUI sinks that never go away. Generally, you can leave those in there, but it might help a marginal flowgraph to turn off the ones you don't need anymore.

4. Close other applications or processes that don't need to be running. These may be slowing down your computer, stealing CPU cycles from GNU Radio. Similarly, rebooting may help.

If all else fails, you may need to upgrade your computer. Installing a faster CPU will typically provide the largest benefit. Having a solid-state drive (SSD) as opposed to a mechanical hard disk will also allow for much faster writes to your hard drive if your flowgraph is streaming large amounts of data to disk.

Gain and SDR Hardware

Besides IQ sampling, another major function SDR hardware carries out is applying *gain*. This isn't the software-implemented, mathematical kind of gain we've discussed in the context of GNU Radio flowgraphs, but rather the result of amplifiers built out of hardware. This is necessary because many radio signals that SDR users try to receive are faint and require a great deal of gain to detect. On the transmit side, SDR users also need gain to boost the strength of the signals they send.

In many SDRs these gains are set using a single value, typically in decibels. In the HackRF, however, the gain occurs in three stages.

The Three Gain Stages

Figure 11-23 shows the three gain stages of the HackRF. These stages are specifically for the receive side of SDR operations. The transmit side goes through similar stages, albeit in reverse.

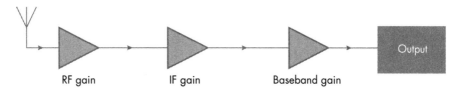

Figure 11-23: The HackRF receive-side amplifiers

The first stage is the *RF gain*, which boosts the raw signal just after it's picked up by the antenna and before any kind of signal processing or sampling. This is a broadband gain stage, meaning there's no filtering prior to it. Rather, the entire RF spectrum supported by the HackRF (1 MHz to 6 GHz) runs through this stage. Another way to say this is that no filtering is done before this gain stage, except the natural filtering that occurs due to the HackRF's limitations. (Nothing has infinite bandwidth.)

There are only two settings for the RF gain: 0 dB or 14 dB. You can enter other values into the RF Gain property of the Soapy HackRF Source block in your receiver flowgraph, but GNU Radio Companion will round your number to the nearest legal value. You need to be careful with the RF gain stage because the higher value could damage your SDR. It takes a very strong signal to do this, much stronger than the FM broadcasts we've previously picked up. However, if you're working with a transmitter situated very close to your SDR's antenna (for example, your transmitter and receiver antennas are on the desk next to each other), damage is a possibility. Keep in mind that this is broadband gain, meaning you don't have to be tuning the SDR to the potentially damaging signal for it to hurt your hardware. For this reason, it's a good idea to set your RF gain to 0 dB when you suspect there might be powerful transmitters close by.

The next gain stage is the *IF gain*. As we hinted earlier, the HackRF, like a traditional fixed-function radio, downshifts the incoming signal to an intermediate frequency before further processing, since there can be engineering benefits to doing so. You don't have to worry about this intermediate frequency at all in your flowgraphs, but nevertheless it exists internally in the hardware, and the signal is further amplified after this downshift occurs.

On the HackRF, the IF gain can be set between 0 dB and 40 dB in increments of 8 dB. Unlike the RF gain stage, this gain is merely amplifying internal signals and can't cause any actual damage to your SDR. If turned up too high, however, it can cause distortion. The HackRF design team recommends that you start with an IF gain of 16 dB and increase if necessary up to the maximum.

The final gain stage is the *baseband (BB) gain*. This stage is applied to both the I and Q signals, just before they enter the ADCs. There's a lot of adjustability here, as the HackRF supports values between 0 dB and 62 dB in 2 dB increments. The official recommendation is to start the BB gain at 16 dB, just like the IF gain, and to increase it in tandem with the IF gain as needed. Once you hit 40 dB and the IF gain maxes out, you can continue to increase the BB gain if necessary.

Notice how each successive gain stage offers both greater gain and increased flexibility as we go from the SDR input to the computer interface. This is because the frequency and bandwidth requirements are lessened as we pass through the different stages, which is a product of the HackRF's use of an intermediate frequency. As mentioned earlier, it's much easier to implement hardware that operates at limited frequencies and over smaller bandwidths.

TRANSMIT GAIN STAGES

For transmission, the HackRF goes through a similar sequence of gain stages as receiving. The difference is that the stages take place in reverse order and there's no BB gain stage, as shown in the following diagram:

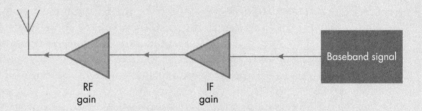

When a signal is output to your SDR hardware for transmission, the digital signal is converted to analog by a digital-to-analog converter (DAC). The resulting analog signal then goes through an IF gain stage at the intermediate frequency. Then it goes through a final RF stage before being passed along to the antenna.

In addition to this hardware gain, it is also possible to control the strength of your transmitted signal by scaling the size of the samples you send to your SDR sink. Inserting a `Multiply Const` block with a Constant of `0.1`, for example, would reduce the output power by a factor of 10, or 10 dB. Just ensure that you don't exceed the maximum specified value for input to your SDR sink, which is typically 1 for both parts of each complex sample.

We'll cover transmit gain in more detail in Chapter 13 when you build your first transmitter.

How to Set the Gain

There are three general principles for setting the gain in your SDR, and juggling them is often a careful balancing act:

- Ensure that your cumulative gain is high enough for the ADCs to convert the analog signal to a sufficiently accurate digital representation.
- Don't apply too much gain at any given stage, or you may cause distortion in the signal.
- Be careful about setting the gain so high that you damage your SDR.

These directives may appear somewhat contradictory; the first point certainly seems to be at odds with the other two. This is why there's something of an art to radio operation.

Based on these principles, you should typically follow a flow like this when you're trying to receive a signal:

1. Set the initial gain to a low level. This will help keep your SDR safe.
2. Run a flowgraph with a QT GUI Frequency Sink block attached to your SDR source. See if you can locate any spikes on the plot that correspond to your target signal. If not, proceed to step 3; if so, skip to step 4.
3. Slowly increase the gain while observing the frequency plot. You should see a peak slowly start to emerge from the noise. If you're using a HackRF, start by increasing the RF gain stage to 14 dB and then move on to the IF and BB stages.
4. Once you've observed a spike corresponding to the target signal, continue to gradually increase the gain. Doing so will increase both the size of the spike and the surrounding noise level, but at first the spike will get bigger rather quickly. Recall that the difference between the signal peak and the noise level is the signal-to-noise ratio (SNR). Continue to increase the gain until the peak stops increasing faster than the noise level. At this point, you can stop. You've maximized the SNR.
5. Demodulate your signal and observe the output, whether this is listening to audio or recovering digital data. If you're experiencing distortion in the sound or the expected digital waveform, try reducing the gain gradually to see if the demodulated signal improves.

If these steps don't work, you likely have an antenna problem (which we'll discuss in the next chapter) or you're trying to receive a signal that's either too weak or too noisy. In other words, the SNR is too low.

How Gain Affects a Signal

You might be wondering why the gain affects the signal peak and noise level the way it does. Perhaps counterintuitively, the goal of applying receive gain isn't solely to make the signal larger. A larger signal by itself doesn't help because the gain affects the noise as well as the target signal. As you turn up the gain, the noise gets worse just as much as the signal gets better.

The purpose of receive gain in an SDR is to help the ADC work better. The maximum input an ADC can measure is called the *full-scale input*. If a given signal is very small relative to the full-scale input, the ADC will capture relatively little information about the signal. For example, imagine an 8-bit ADC with a full-scale input ranging from −1.0 to +1.0. Such an ADC would produce $2^8 = 256$ different output values, with its maximum output corresponding to +1.0 and its minimum corresponding to −1.0. Now imagine a very small input signal is applied to this ADC, as shown in Figure 11-24.

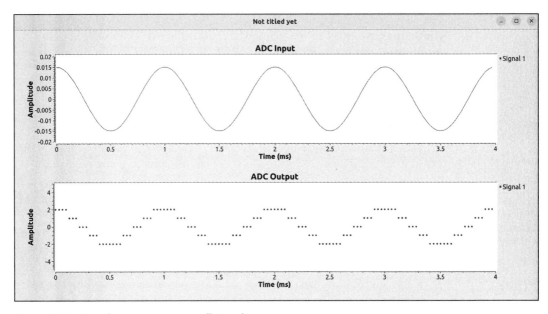

Figure 11-24: Sampling a very, very small signal

The very small sinusoid in the figure only ranges from +0.015 to −0.015. Because its range is so narrow, the sinusoid is converted to a relatively small number of digital values; it's all just −2s, −1s, 0s, 1s, and 2s. It may be an 8-bit ADC on paper, but it's producing only 5 different measurements out of the 256 possibilities, so its practical resolution is only 2 or 3 bits. With such a low effective resolution, the ADC can't accurately measure the signal or provide much information about it. This measurement error manifests as noise, resulting in a worsened SNR.

On the other hand, if the signal is large enough to be close to the ADC's full scale, the conversion produces values like those shown in Figure 11-25.

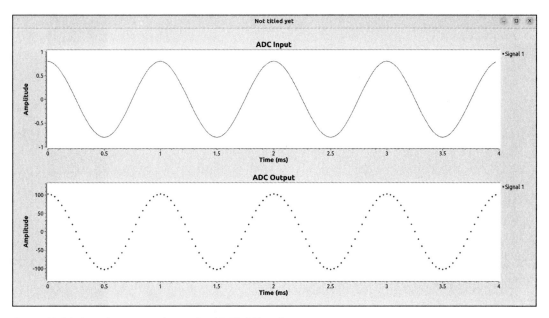

Figure 11-25: Sampling a signal near the ADC's full scale

In this case, the analog signal ranges from +0.8 to −0.8, or 80 percent of the full range. As such, the ADC is able to use a much greater number of values to describe the signal, providing significantly more detail. The measurement error is much smaller than the signal values, and the resulting SNR can be much better.

If you turn up the receive gain too high, however, another problem can occur: *clipping*. This is when the gain is so high that it produces an ADC input greater than full scale. Figure 11-26 shows an example of clipping, where the input exceeds the maximum and minimum full-range inputs of +1.0 and −1.0.

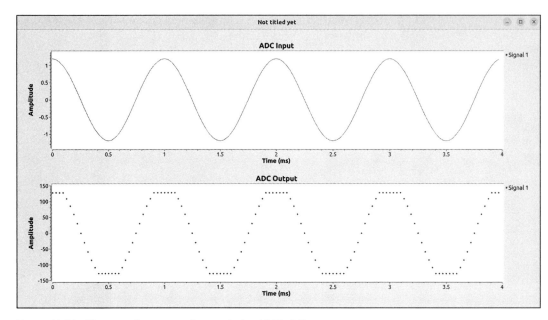

Figure 11-26: Clipping when a signal exceeds the ADC's full range

The ADC samples are correct until the input signal reaches its uppermost values, at which point the ADC can't output any higher measurements. (The same thing happens on the low end.) Lacking any other options, the ADC outputs the maximum (or minimum) value it can, which is no longer a meaningful measurement of the signal. This introduces distortion into the ADC conversion values and can prevent you from recovering a faithful representation of the original signal.

A Better SDR Model

With everything we've discussed so far in this chapter, we have all the information we need to draw a more detailed model of how SDR hardware works. Figure 11-27 shows a block diagram of the HackRF board we've been using for this book. There may be slight variations on the gain stages for other SDR designs, but overall, the structure should largely be similar.

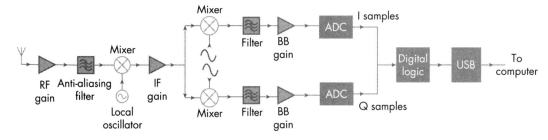

Figure 11-27: A more detailed model of the HackRF SDR

The signal from the antenna feeds directly into the RF gain stage. This in turn is downshifted to the intermediate frequency. This IF signal is then amplified by the IF gain stage before the I and Q signals are generated by a pair of sinusoids phase-shifted 90 degrees from one another. The resulting I and Q signals are low-pass filtered to prevent aliasing and then amplified by a third BB gain stage. Finally, these filtered and amplified I and Q signals are sent to a pair of ADCs for conversion.

A bit of digital computation is still required to transform the raw ADC output into a format that can be sent over the USB port. Once that's done, the HackRF driver on the host PC sends a complex number stream to the Soapy HackRF Source in GNU Radio for you to process to your heart's content.

It's important to stress that all of this is going on under the hood of your SDR hardware, without your intervention. You don't have to worry too much about technical details behind the hardware to use it as a basic receiver or transmitter. However, understanding your hardware's inner workings may be helpful as you work on more sophisticated SDR projects.

DC Offset

We've discussed how SDR hardware works, but there's one last aspect of your hardware that you need to be aware of: *DC offset*. This is a false spike that can occur exactly in the middle of a receiver's frequency plot. In the hardware FM receiver flowgraph from Chapter 9, for example, this spike was fairly pronounced. It's highlighted in Figure 11-28.

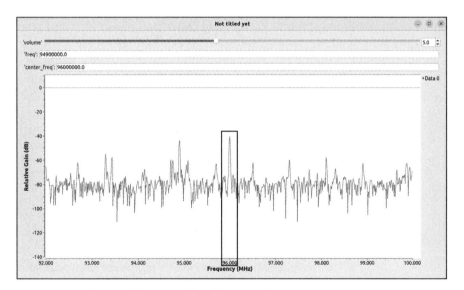

Figure 11-28: The DC spike in a HackRF frequency plot

The spike is caused by a real physical phenomenon, but it isn't related to the RF signals you're trying to capture. It occurs at 0 Hz regardless of the RF input and is due to imperfections in the SDR's amplifiers, filter, and ADC circuitry, along with a mathematical quirk in the FFT computation.

In a perfect world, if you applied a static 0 V input to an amplifier circuit with a gain of 10, you'd see 0 V on the output, as shown in Figure 11-29. Multiply 0 by 10 and you still get 0.

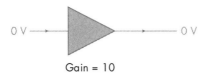

Gain = 10

Figure 11-29: An ideal amplifier

In a real-world amplifier, however, you'll see an output slightly different from 0. Imperfections in the amplifier circuitry create an *offset error*, in which the input signal is interpreted as slightly higher or lower than it actually is. Let's say this offset error turns the incoming 0 V signal into 0.01 V. Then the amplifier multiples this 0.01 V signal by 10, so the offset error causes an output voltage of 0.1 V, as shown in Figure 11-30.

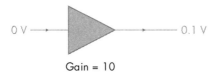

Gain = 10

Figure 11-30: A nonideal amplifier with an input of 0 V

Even if the input signal is nonzero, you'll see the same offset error relative to what the output should be. In Figure 11-31, for example, the input is 0.2 V, but rather than the ideal 2.0 V output, the 0.01 V input offset causes the output to be 0.1 V higher, at 2.1 V.

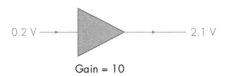

Gain = 10

Figure 11-31: A nonideal amplifier with an input of 0.2 V

Strictly speaking, engineers differentiate between different types of offset more formally than we've done here, but that's deeper than we need to go. The key factor for SDRs is that the primary offset looks like a constant value that's been added to all your signals. Engineers call an electrical quantity that doesn't change over time a *direct current (DC) quantity*, so this offset effect is called a *DC offset*.

What's the frequency of a DC, or constant, voltage? Since it doesn't change, its frequency is 0 cycles per second. This is why you have the spike at 0 Hz, right in the center of your frequency plot, between the negative and positive frequencies. This is often referred to as the *DC spike*.

The DC spike doesn't represent a legitimate RF signal, and it will distort any signal that's located at 0 Hz, so you need to work around it. In practice, it's best to avoid setting your SDR's center frequency right at the frequency of the signal you're trying to capture. Instead, you should set the center frequency to one side or the other. In Figure 11-32, for example, the target signal is at 433.9 MHz, but the center frequency has been set to 433.5 MHz. This is called *offset tuning*.

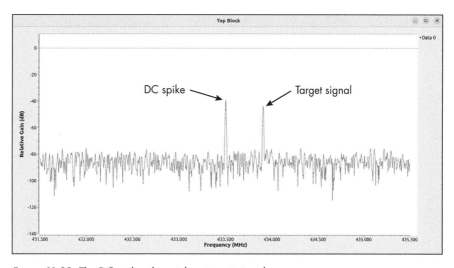

Figure 11-32: The DC spike alongside a target signal

Although the DC offset is involved in all this, as it's creating the spike we're trying to avoid, that's not the offset referenced by "offset tuning." Rather, the term refers to a *frequency offset*, which is the difference between the SDR center frequency and the frequency to which you want your receiver to tune. Another way to describe this is the difference between the hardware frequency (center_freq) and the software tuning frequency (freq).

There are more complicated methods to avoid the DC spike, such as using calibration to minimize the offset error, but it's usually easier to just move the spike out of the way, frequency-wise. Up to now, all our radio receiver flowgraphs have done this by using offset tuning; we just didn't point out that it was happening. For example, the AM receiver flowgraph from Chapter 4, shown again in Figure 11-33, performs offset tuning using the complex sinusoidal multiplication highlighted by the rectangle labeled "Shifter." This multiplication ensures that input RF data will be shifted by 20 kHz, centering the target signal rather than the DC spike. The target signal then goes through the subsequent low-pass filter, while the spike is eliminated.

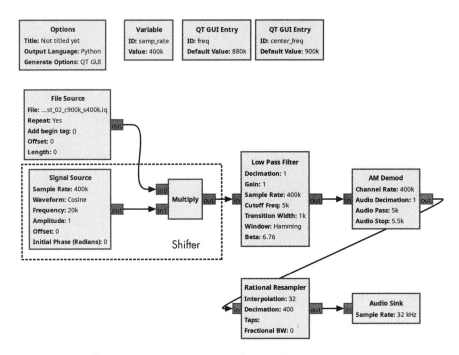

Figure 11-33: Offset tuning in the AM receiver flowgraph

The FM hardware receiver flowgraph from Chapter 9, shown again in Figure 11-34, performed the same operation but with a different block, Frequency Xlating FIR Filter.

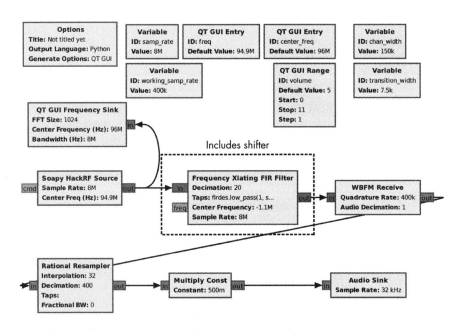

Figure 11-34: Offset tuning in the FM hardware receiver flowgraph

When working with IQ data from a File Source, it's crucial to have a fixed value for your center_freq variable corresponding to the center frequency to which the SDR was set when it originally captured the data. (This value has been embedded in the names of our IQ files, immediately following the *c* character.) If you have an incorrect center_freq value, your flowgraph will present an incorrect interpretation of the frequencies of the signals in the IQ data.

If you're working with live IQ data from an SDR, you can set your center _freq value in a more useful way. Open *ch_11/fm_hackrf.grc*, which is the same FM hardware flowgraph we built in Chapter 9 and is shown in Figure 11-34. Then change the value of center_freq to freq - samp_rate/4. This ensures a center frequency value that's offset from your desired tuning frequency by one-quarter of the sample rate, as shown in Figure 11-35.

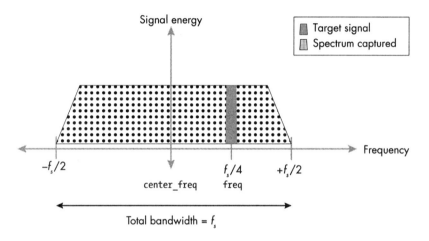

Figure 11-35: The relationship between SDR settings and tuning values

Notice the position of the software tuning frequency (freq) within the capture range, halfway between the center of the capture window and the Nyquist frequency. Computing your SDR's center frequency in this way forces the DC spike out of the way of your signal, guaranteeing that freq will be a legal value within the capture range but one that doesn't fall on the spike. Because of the automatic computation, you can now change your tuning frequency (freq) to any other value supported by your SDR's frequency range, and the center frequency will automatically recompute to support it.

Important SDR Specs

Now that you're equipped with a general understanding of how SDRs work, let's look at the specifications commonly listed for them. We'll tell you what to look for in an SDR and how the specs affect what you can do:

Operating frequency The range of frequencies that you can receive and transmit with your SDR. With respect to GNU Radio, this is the range of legal values that you can enter for the Center Frequency property of your SDR source or sink. For the HackRF One, this range goes from 1 MHz to 6 GHz, meaning the HackRF can receive and transmit along the entirety of the HF, VHF, and UHF bands, as well as parts of the MF band on the low end and the SHF band on the high end. The low-end limitation means you'll have trouble receiving about half of the AM broadcast band, which ranges from around 500 kHz to 1.6 MHz. The high-end limitation will impact you if you plan on working with satellite transmissions or some 5G cellular communications. Impressively, the HackRF and many other SDRs will operate at all 3G and 4G telephony frequencies, as well as many Wi-Fi frequencies.

Full duplex or half duplex   Describes whether the SDR can transmit and receive at the same time. If you're using the HackRF One, a half-duplex device, you must choose to either transmit or receive at any given time. If you'd like to do both, you must either switch from one mode to the other or use two HackRF devices. Full-duplex devices, which can transmit and receive simultaneously, tend to be more expensive.

Number of channels   Many higher-end SDRs contain multiple full-duplex channels, allowing you to implement multiple transmitters and receivers that can operate simultaneously.

Sampling rate   Defines what range of sampling rates you can use. In some cases, it even specifies the sampling rates you must use, rather than a continuous range. The HackRF One's ADCs are specified to operate between 2 Msps and 20 Msps. The device will also work at rates lower than 2 Msps, but you may get some aliasing from adjacent signals due to analog effects that you can't fix in your flowgraph. Best practice is to sample at a minimum of 2 Msps and immediately decimate if you'd like a lower rate.

Resolution   The number of bits produced by the ADCs on the receive side or the number of bits utilized by the DACs on the transmit side. When you sample a signal, there's always some amount of error converting the analog signal to a digital value. The HackRF has an 8-bit resolution, meaning it converts each analog value it sees into one of 256 (2^8) different digital values. A higher-performing SDR may have a resolution of 12, meaning it will produce one of 4,096 (2^{12}) different digital values. Clearly the latter will provide a more accurate representation of the original analog signal. Because the inaccuracy is random, the error that does occur looks just like noise. The formal term for this error is therefore *quantization noise*. Higher resolution means less of this kind of noise.

RX gain   The receive-side gain we talked about earlier in the chapter. There will be some maximum value to which this can be set in the SDR source block in GNU Radio. Your SDR will likely have a single gain control, but it may consist of multiple gain stages, as it is for the HackRF One.

TX gain   The transmit-side gain, which will also have a maximum value and may consist of more than one gain stage.

RF output power This will likely be a set of different values depending on the frequency at which you're transmitting. The HackRF One, for example, can produce 30 mW of transmission power at lower frequencies, but this slowly drops as frequency increases. By the time you get to 4 GHz, the maximum power drops to less than 1 mW.

System power Your SDR may have a specific power consumption given in watts or a specific supply current given in milliamperes (mA). Most often, though, what you care about is whether you can power your SDR from the USB bus without an external power adapter. With most affordable SDRs like HackRF devices, the answer is yes.

Interface Many SDRs will interface to your computer via USB. Some more expensive devices, however, use an Ethernet connection (1 GB/s or 10 GB/s). When attached to a network, these Ethernet-enabled SDRs can be accessed by multiple computers a great distance away. Other expensive SDRs use PCIe or M.2 connectors, so they must be plugged directly into the motherboard of a desktop PC or specially designed computer.

Antenna connector Typically an SMA connector, as on the HackRF board. Some other SDRs, like certain RTL-SDR dongles, use an MCX connector. We'll discuss RF connectors more in the next chapter.

Clock input This port can be used to supply a better clock signal to your SDR than the one the SDR generates on its own. By *better*, we mean more accurate in an absolute sense, as well as more stable over time and with temperature changes. We won't be needing this feature for any of the exercises in this book.

Clock output When operating multiple SDRs simultaneously to implement a complex communication system, it can be helpful to have them running off the same clock. Connecting the CLKOUT port of one SDR to the CLKIN of another will synchronize the two devices.

Although all of these specifications are important, the most important for lower-priced SDRs are operating frequency range and sampling rate. As you climb in SDR price, also look for better resolution ADCs/DACs and additional channels.

Conclusion

In this chapter, we examined the inner workings of your SDR hardware. We talked about IQ sampling, the technique the hardware uses to convert radio signals to digital data. You saw how an SDR's capture window has a bandwidth equal to its sample rate and centered at its center frequency, and you also learned how to set the receive gain for your SDR and avoid overrun errors. Finally, we walked you through several definitions that will be useful when you evaluate SDR hardware for use or acquisition.

Our look at SDR hardware isn't complete, however; in the next chapter we'll consider antennas, connectors, and other peripheral devices needed for SDR systems.

12

PERIPHERAL HARDWARE

You've learned enough to start using your SDR with GNU Radio, but the more radio frequency (RF) work you do, the more you'll have to think not just about your SDR itself but also about peripheral hardware like your antenna, connectors, and so on. In this chapter we'll outline the peripheral devices you need to be most effective in your SDR use. The chapter isn't meant to be exhaustive but should equip you to handle most basic and intermediate projects. We'll also discuss overlooked factors that can affect your SDR's performance, such as the specifications of your computer itself, and we'll look at some hardware-based solutions for reducing noise in your radio systems.

Antennas

The most important SDR peripheral is your antenna. You've already been using an antenna, the ANT500, in previous chapters, but we haven't spent any time explaining what it's doing and why it works. Perhaps you have an intuitive understanding that an antenna is a device that transmits or picks up signals from the air. If you're old enough, you might even have some practical experience manually tuning an antenna, though you probably didn't think too much of it, as you were fiddling with rabbit-eared antennas to improve the picture on your favorite 1980s television show featuring a Hawaiian shirt–clad private investigator. Now let's turn that general intuition into practical knowledge you can apply to your SDR experiments.

First, though, a familiar disclaimer: this isn't a physics textbook. There won't be a deep dive into electromagnetic field theory, and we won't be deriving Maxwell's equations. That would be further beyond the scope of this book than even the topic of IQ sampling. The main goal is to help you understand a few basic characteristics of a handful of different antenna types. We'll be applying the lightest of scratches to the surface here, but enough for you to get a lot of basic work done.

We'll start with a little bit of physics, specifically electric fields and magnetic fields. You've encountered both in your daily life. Ever felt the hairs stand up on your arm when near a static-riddled blanket or other fabric? That's an electric field at work. Meanwhile, refrigerator magnets are probably one of many interactions you've had with magnetic fields. But we don't want to consider electric and magnetic fields in isolation; the key for radio is in how the two interact.

A couple of interesting things happen when you create an electric field across a wire, for instance, by connecting a battery. First, it causes an electric current to flow. Second, the moving charges that make up that current generate a magnetic field. Another interesting thing happens when you place a magnet near a wire and then move it around. The resulting changes in the magnetic field generate an electric field in the wire. In other words, electric changes trigger magnetic changes and vice versa!

A *very* interesting thing happens when you run a changing current through a specially designed wire. The moving electrical charge in the wire creates a changing magnetic field. That changing magnetic field in turn generates a changing electric field, and so on ad infinitum. This is called an *electromagnetic wave*, and radio waves are a specific type of electromagnetic wave. Such waves will propagate (we've heard that term before) away from the wire at the speed of light.

An antenna is nothing more than that "specially designed wire" just mentioned. There are a few different antenna characteristics and designs, and we'll go through those at a high level in the next couple of sections. Before we move on, though, consider another question: What do you think happens when we run a changing current through a wire that hasn't been specially designed as an antenna? In fact, such a wire is also an antenna, just a relatively bad one. This means that any wires carrying changing currents (which is almost all the wires in a powered-up electronic system) will be sending out electromagnetic waves in all sorts of random directions. What does this mean for us SDR practitioners? More noise. Like we didn't have enough already.

Characteristics

You need to consider several different factors when selecting an antenna. The most important are frequency, bandwidth, gain, and directivity. We'll also say a word about impedance, but for the most common SDR antennas, this will be a straightforward issue.

Frequency

Antennas are typically designed for a specific range of frequencies. The inconvenient thing that physics has mandated, however, is that an antenna's physical size is directly proportional to the wavelength, or inversely proportional to the frequency, for which it is designed. A general rule of thumb is that most antennas will be about one-fourth to one-half the size of the wavelength for which they're designed.

Clever design can shrink this size, and if the signal is particularly strong, an optimal design is unnecessary. For example, the wavelength of FM broadcast signals is about 3 meters. But did you notice that you were able to pick up FM signals regardless of whether your telescoping antenna was extended to one-fourth wavelength (about 2.5 feet)? FM broadcast signals are so powerful that they can often be picked up with a suboptimal antenna. Still, the general rule is this: the higher the frequency, the smaller the antenna; the lower the frequency, the larger the antenna. This holds for both receiving and transmitting.

Bandwidth

Antennas can be designed with wider or narrower bandwidths. There are advantages and disadvantages to each. If you have a wider bandwidth, you can receive a greater variety of signals. If you have a narrower bandwidth, though, you can better ignore any noise or unwanted signals coming in on frequencies outside your narrow bandwidth. You can think of an antenna's bandwidth as a band-pass filter on the very front end of your receiver or at the final stage of your transmitter.

Gain

Larger and more complex antenna designs will pick up a stronger signal than smaller and simpler ones. We express this difference by attributing gain to the antenna. This gain is measured in *dBi*, which stands for "decibels relative to an isotropic antenna." You can think of an isotropic antenna as a standard reference antenna that radiates equally in every direction.

You can't simply add the antenna gain in dBi to the rest of your radio's gain stages (in dB) to compute a total system gain. Instead, dBi allow you to determine how much gain you'll be adding to a system by using one antenna versus another. Simply compare the antennas' dBi values to each other.

Directivity

If you have a simple, straight antenna pointing upward, it will radiate equally in all horizontal directions as well as pick up signals from any horizontal direction. This is why straight antennas are called *omnidirectional*. The term can be a bit misleading: while an omnidirectional antenna works equally well all 360 degrees around the plane perpendicular to where it's pointing, it doesn't necessarily work so well in the direction the antenna is actually pointing (vertically, in this example). Whatever direction you point the antenna in, its optimal plane of operation will be perpendicular to that direction, as shown in Figure 12-1.

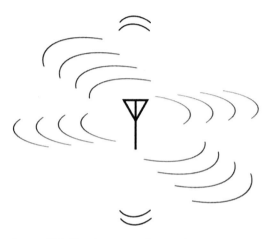

Figure 12-1: The directivity of a simple antenna

Other types of antennas are more directional, meaning they'll radiate more strongly in particular directions (for example, the direction in which they're pointing). On the receive side, these directional antennas will pick up signals more readily from some directions than others. Meanwhile, some antenna radiation and reception patterns are more complicated than simply "everywhere" or "point at the target." We'll explore this further when we look at specific antennas later in the chapter.

Impedance

Impedance is a term describing how hard it is to push electrical current through something. An antenna's impedance therefore tells us how hard it is to push current through the antenna. Since we're mostly concerned with off-the-shelf antennas and relatively short cable lengths, we don't need to spend a lot of time with impedance. The issue becomes more of a concern if you're trying to build your own antenna or use a more unusual antenna type.

That said, you *do* need to make sure the impedance of your off-the-shelf antenna matches the impedance of your SDR's antenna port. Otherwise, you'll lose a lot of power going in and out of your SDR, which will reduce its effectiveness at both receiving and transmitting. For the HackRF One, as well as most SDRs, the optimal impedance is 50 ohms (Ω). Fortunately, many antennas are designed for a 50 Ω impedance, so if you want to use a specific antenna type, this shouldn't be a major factor.

Types

There's an enormous variety of antenna designs out there, as a quick internet search will show you. In this section, we'll consider a few basic types, which should cover most of the situations you'll encounter. If you'd like a more in-depth look at this topic, a web search of ham radio antennas will be informative.

Whip and Telescoping

The simplest antenna type is probably a *whip antenna*, sometimes called a *monopole*. It consists of a fixed length of straight conducting material, often flexible and sometimes covered in a protective coating. An example is shown in Figure 12-2.

Figure 12-2: A whip antenna

Because of their fixed length, whip antennas are optimized to a particular frequency. They have a fairly wide bandwidth and are omnidirectional, so you don't need to worry about aiming them at your target. Because they transmit and receive best in the perpendicular direction, however, it is usually best to orient them vertically. These whip antennas have moderate gain, roughly 2 dBi in perpendicular directions (about 1.6 times greater than the anisotropic antenna).

Telescoping antennas, such as the ANT500 shown in Figure 12-3, are simply variable-length monopole antennas.

Figure 12-3: Telescoping antennas, retracted (left) and extended (right)

You tune a telescoping antenna by extending or retracting it to match the one-quarter wavelength of the desired operating frequency. In other respects, telescoping antennas behave like whip antennas.

Dipole

Dipole antennas, like the one shown in Figure 12-4, have a T shape. This shape causes them to have significantly different characteristics than the previous types. First, the bandwidth is typically much smaller than for a monopole. This is good if your antenna is matched with your target frequency, but it means you'll have less flexibility to handle situations where the target frequency is different.

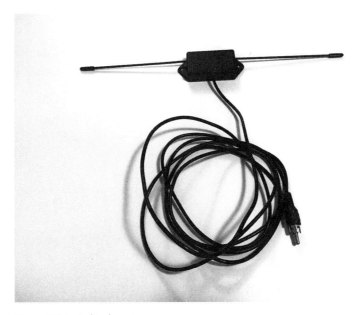

Figure 12-4: A dipole antenna

Most dipole antennas that you encounter will be either a quarter-wave or half-wave. This means each of the two crossbar portions will be either a quarter-wavelength or a half-wavelength in size. If such a dipole antenna is oriented horizontally, meaning the crossbar of the T is parallel to the ground, then the antenna will behave directionally. This orientation is shown in Figure 12-5.

Figure 12-5: A dipole antenna parallel to the ground

An optimal horizontal orientation will produce about 5 dBi greater gain, or 3 dB greater than an equivalently sized monopole, but only in the directions perpendicular to the antenna. Signals coming from (or going to) a direction parallel to the crossbar of the T will have significantly less gain than an equivalent monopole. Figure 12-6 shows the gain pattern.

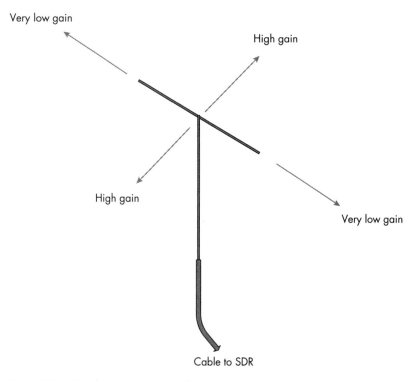

Figure 12-6: Dipole antenna gains relative to orientation

The dipole essentially sacrifices gain to the sides in favor of gain to the front and back. Because you'll also be getting less noise from those sideways directions, you may experience a better signal-to-noise ratio with a dipole than with a monopole.

Loop

Loop antennas are typically very compact, even for long wavelengths. They're often used when the equivalent monopole or dipole antenna would be unreasonably large. In fact, nearly all AM radio receiver antennas are of this type, such as the one shown in Figure 12-7.

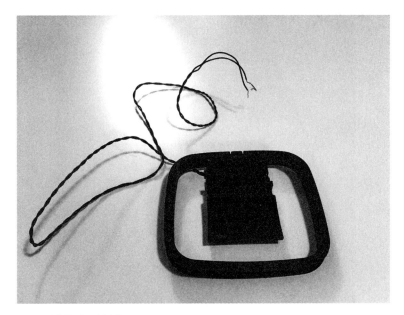

Figure 12-7: An AM loop antenna

Loop antennas are directional, receiving signals best along the axis of the coil, when the loop is facing the target, as shown in Figure 12-8. Small loop antennas aren't suitable for transmission in most cases.

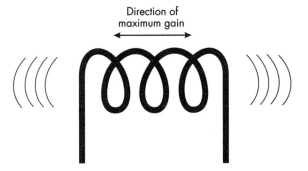

Direction of
maximum gain

Figure 12-8: Loop antenna directivity

Other

There are a host of more elaborate antenna designs, each with its own combination of bandwidth, directionality, and gain. Perhaps the most common of these other designs is the *yagi antenna*, such as the one shown in Figure 12-9.

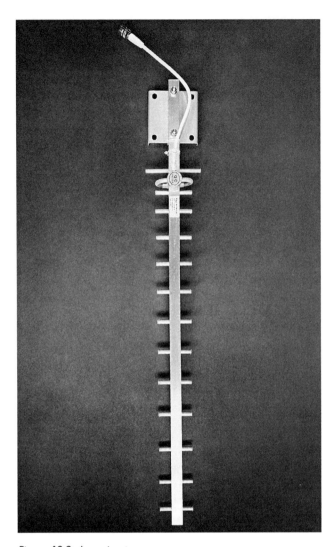

Figure 12-9: A yagi antenna

Yagis are complex assemblies of many pieces, resulting in a highly directional, high-gain antenna. If they look familiar, it's because they used to be in common use on rooftops for television reception.

If you search around a bit for antennas, you'll also encounter something called an *active antenna*. These antennas can use any kind of physical topology but integrate an amplifier to provide extra gain for receiving signals. These antennas require a supply voltage to power the amplifier, which can be applied externally or through the antenna connector itself. The HackRF One provides 50 mA of current to power these active antennas, but not all SDRs include this functionality.

Polarization

For several of the antennas in the last section, we talked a bit about how the direction in which you point your antenna often matters. It turns out that the rotational orientation of your antenna also matters. If you imagine standing right behind the receiver, we're talking about the rotation of the transmitting antenna from your visual perspective. This is because radio waves are *polarized*, meaning that the electromagnetic wave has its electric and magnetic components each on specific axes. Again, imagine a radio wave shooting right at you. The electric component of the wave will always be oriented perpendicularly to the magnetic component. (To review what this looks like, see the image in "Wavelength vs. Frequency" on page 167.)

The specific tilt of a transmitting antenna will determine the polarization of the resulting radio waves. If a receiving antenna is oriented in the same way as the transmitting antenna, it will have maximum gain. If one antenna is rotated with respect to the other, gain will suffer to the extent that the orientations are different. In theory, a receiving antenna 90 degrees out of orientation from the transmitting antenna won't sense the transmitted signal at all. In practice, though, this doesn't happen. The orientations will never be exactly out of alignment, and waves scattered off intervening objects will also muddy the polarization, rendering it imperfect.

Ultimately, polarization means about the same thing to you that it meant to kids in the 1980s tuning their televisions: if your received signal is too weak, you may improve it by tweaking the rotation of your antenna.

An Antenna Experiment

Now that you know a bit about how antennas work, let's see what happens when you make changes to your antenna while running your FM radio project from Chapter 9. First, make sure all your SDR hardware is connected properly. The ANT500 should be screwed into the ANTENNA port of the HackRF One, which should in turn be connected via USB to the computer running GNU Radio Companion. Once you've verified this, open and run *ch_12/fm_hackrf.grc*. This flowgraph is identical to those we used previously for FM reception.

Try retracting the antenna all the way to its shortest state, then tune the radio to the strongest signal you see by entering its frequency into the freq variable. In the example shown in Figure 12-10, we've set it to 94.9 MHz, corresponding to the highest peak on our frequency plot. Your strongest peak will likely be different.

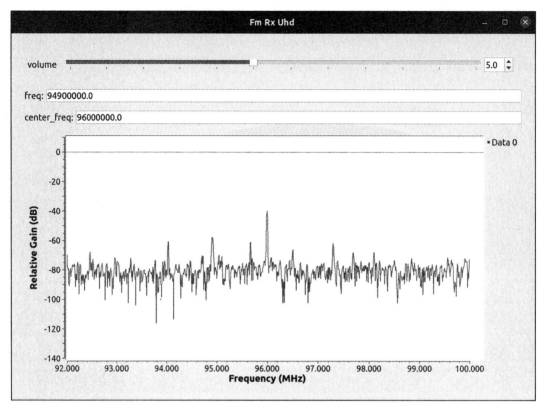

Figure 12-10: The FM receiver flowgraph output with the antenna retracted

Take note of the signal strength by hovering your cursor over the dotted green horizontal line. It moves around, so just try to get it in the middle of its jittering. On our plot it was around −58 dB. Now extend the antenna by one segment and watch what happens to the signal strength. In our testing, as shown in Figure 12-11, the size of the peak changes.

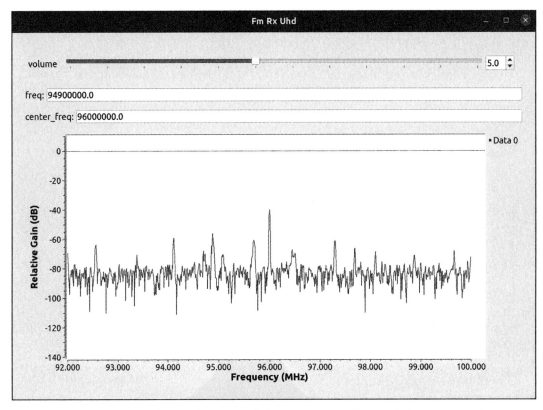

Figure 12-11: The FM receiver flowgraph output with the antenna partially extended

In this case, the power increased very slightly to about –55 dB. Now go ahead and extend the antenna all the way to its maximum length. Our results are shown in Figure 12-12.

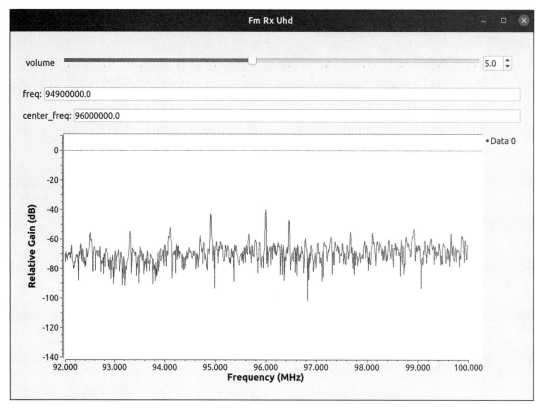

Figure 12-12: The FM receiver flowgraph with the antenna fully extended

You'll likely see the signal strength jump up quite a bit. On our rig it went nearly to −40 dB. This correlates quite well with the claim we made earlier that antennas typically operate better when sized to a quarter or half the target wavelength.

What's the wavelength of an FM broadcast signal? As discussed in Chapter 8, we calculate this by dividing the speed of light by the signal's frequency. One quick way to do the math is to recognize that the speed of light, 3×10^8 meters per second, can also be written as 300×10^6, or 300 million meters per second. The frequency of a typical FM radio signal is about 100 MHz, or 100 million cycles per second. The wavelength of an FM broadcast signal is then 300 million meters per second divided by roughly 100 million cycles per second, which yields 3 meters per cycle. One-quarter of 3 meters is about 29.5 inches. Fully extended, the ANT is just over 33 inches, very close to a quarter wavelength.

You can also tweak the antenna more precisely if you do the computation for the exact frequency you want. In our case, that would be 300 divided by 94.9; your station will probably be different. The quarter wavelength for our station's frequency is 31.1 inches. Indeed, when we shorten the antenna slightly, to just over 31 inches, the signal is boosted a tiny bit more, increasing by about 0.5 dB.

Take a minute to consider the magnitude of the changes you've witnessed here. The poorly designed, excessively short antenna we started with gave us about −58 dB of signal strength, while a more optimized antenna gave −40 dB. Simply extending the antenna length provided an 18 dB improvement. That's more than 10 times greater, as you may recall from our discussion of decibels in Chapter 5.

How Computers Affect SDRs

One piece of peripheral SDR hardware is so obvious that you might forget to consider it: the computer itself. Given that we're talking about *software-defined* radios, it's only natural that the software will be affected by the hardware on which it runs. The following are the key computing concerns:

Computation speed   If your computer is too slow, you'll experience underflow and overflow conditions, compromising your ability to send and receive signals. You can often mitigate these problems by optimizing the performance of your flowgraph.

Interface speed   If your USB or Ethernet interface can't support the necessary transfer rates, you'll again have overflow and underflow issues. These can only be fixed by addressing interface congestion or by reducing the sample rate of the SDR, thereby reducing the flowgraph's data rate. Be careful when using a device with a USB 3 interface that you don't connect it to a USB 2 port, or you will likely face these issues. You may also run into interface speed issues when running your software in a virtual machine, as the virtual machine software drivers may not be able to keep up with the data speeds your SDR requires.

Storage capacity   When working with SDRs, we often capture data and store it to a file for later analysis. Raw IQ files can grow very large, however, easily climbing to multiple gigabytes in size. If your computer is equipped only with a small drive, we recommend purchasing additional storage. This could be either a second solid-state drive (SSD), a mechanical rotating hard drive, a USB thumb drive, or even an SD card. Whichever media you choose, take care to ensure that the transfer speed of the drive is higher than the rate of data you intend to capture.

Noise   Your computer will inflict noise upon your SDR in two primary ways. First, you may experience power supply noise, which is carried through the USB cable to your SDR. Since most SDRs are powered by the USB bus, this effect is difficult to eliminate entirely. Second, the computer will radiate electromagnetic interference (EMI), which will be picked up by the SDR.

We'll look at the methods for minimizing your computer's noise impact, as well as other kinds of noise, in the next section.

Mitigating Noise

We've discussed noise in previous chapters, and you know it's a bad thing, but what do you do about it? Here are several adjustments you can make to your peripheral hardware to reduce the noise getting into your system. Most of these tips focus on the receive side, but a few are relevant to transmission as well:

Antenna orientation   If you have a directional antenna like a dipole, the direction the antenna is pointing can make a big difference. See if you can find an orientation that reduces your noise level while still pointing the high-gain surface of the antenna toward your target signal. The goal is to orient the antenna such that the SNR is maximized.

Power source   The power supplies of most computers are very noisy, especially those of desktop computers. This can also be problematic with cheap, third-party laptop chargers. If you're using a laptop, you may see an improvement in your noise levels when you unplug your charger; battery-powered hardware is typically the lowest-noise option.

Shielded cables   If you're running a length of cable between your antenna and SDR instead of attaching the antenna directly, be sure to use a shielded cable. This essentially provides a layer of metallic material near the outside of the cable that prevents the inside from picking up noise.

Ferrite chokes   To reduce any power supply noise that may be creeping up your USB cable, use *ferrite chokes*. These are essentially filters that can be added to a cable externally. You can either buy a cable with chokes built into each end, as shown on the left of Figure 12-13, or buy the chokes separately and clamp them onto any cable, as shown on the right. Either option is relatively inexpensive.

Figure 12-13: A USB cable with a built-in ferrite choke (left) and one with a separate ferrite choke attached (right)

Beyond these measures, one of the simplest things you can do to reduce noise is to just move your SDR. Because the noise radiated by a nearby noise source will drop off rapidly with distance, moving your radio away from it will often improve matters. To see this effect in action, try executing a simple flowgraph that directs the SDR output into a QT GUI sink and watch the noise levels change in real time as you move the SDR around.

Connectors

Another piece of peripheral hardware to consider is the connector linking your antenna to your SDR. There's a dizzying variety of different connector types out there. If you're using the HackRF One and the ANT500, you don't need to worry about it, as they're already compatible. If at some point, however, you decide to start working with different antennas, you may have problems plugging them into your SDR.

In this section, we'll survey the main types of connectors you may encounter. First, though, it's important to highlight the distinction between *male* and *female* connectors. With nearly every connector type in existence, a proper connection is made when a male connector is inserted or screwed into a female connector. When seeking out a connector or adapter, make sure that you're connecting not only to the right type but also to the correct gender.

SMA   This is a common connector type for low-power RF devices. The HackRF One uses a female SMA connector to attach to an antenna with a male SMA connector. A connection is made by inserting and then rotating the antenna-side connector in a manner similar to the coax cable you use for your television. Both male and female SMA connectors are shown in Figure 12-14.

Figure 12-14: Male (left) and female (right) SMA connectors

MCX   This is a much smaller connector type used by some RTL-SDR dongles, as well as other SDRs. To attach, you simply plug it in, with no rotation required. Figure 12-15 shows both genders of the MCX connector.

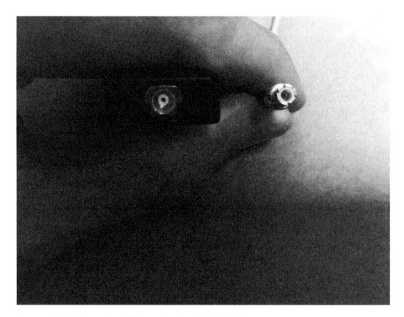

Figure 12-15: Female (left) and male (right) MCX connectors

RCA You may have seen these before in home entertainment systems, shown in Figure 12-16, but some antennas use RCA connectors as well. This is another simple plug-in connector with no rotation needed.

Figure 12-16: Male (top) and female (bottom) RCA connectors

Coaxial These are the same as the connectors your television likely uses for cable or an antenna. As these are threaded connectors, you must insert and then rotate. Figure 12-17 shows male and female coax connectors.

Figure 12-17: Female (left) and male (right) coax connectors

BNC Typically used for more expensive RF equipment, BNC connectors also have an insert-and-rotate mechanism. Instead of a threaded screw requiring many rotations, however, the BNC connector locks into place after a half-turn. Both connectors are shown in Figure 12-18.

Figure 12-18: Female (left) and male (right) BNC connectors

SO-239 This is another type of insert-and-rotate connector, similar to coax. Figure 12-19 shows an example.

Figure 12-19: A female
SO-239 connector

RP-SMA   There's one final connector type worth mentioning, one with
an odd look and an odd history. Thanks to the Federal Communications
Commission (FCC) not wanting you to remove the antenna from your
Wi-Fi router and replace it with a different one, they dictated that all
Wi-Fi devices sold in the United States cannot be compatible with the
standard SMA form factor. Thus was born the *RP-SMA* connector, standing for *reverse-polarity SMA*. This connector hybridizes the male and
female aspects of the original SMA connector. One side contains the
male central plug and the female inner threaded sleeve, while the other
side has a female central receptacle and a male outer threaded sleeve.
Figure 12-20 shows a female RP-SMA connector.

Figure 12-20: A female
RP-SMA connector

When purchasing SMA hardware, be careful to verify it's true SMA
and not RP-SMA. If you *do* need to connect RP-SMA, or any other type of
non-SMA connector, you can always use an adapter, as we'll discuss in the
next section.

Building an SDR Toolkit

We've considered several different types of SDR peripherals, but you definitely don't need to rush out and get them all right now. The longer you
work with SDRs, however, the more you'll appreciate having access to a
toolkit filled with useful hardware. The following suggested items are listed
roughly in priority order, though you'll need to tailor your toolkit to your

own requirements. Hopefully this section will get you thinking about what you'll need. Fortunately, most of the peripherals discussed are cheap and easy to find online.

Antennas

You can get a lot of mileage out of the ANT500 that likely came with your HackRF One. Because you can adjust the length of the antenna to tune it to a wide variety of frequencies, it's one of the best general-purpose antennas out there.

If you plan on working with specific RF bands and need greater directivity or gain, you should also acquire a dipole antenna designed for your target band. However, dipoles become quite large as the frequency drops, so they'll become progressively less portable below 100 MHz.

If you need to work in kHz frequencies, plan on getting a loop antenna for receiving. Since these are simply coils of wire with a radius determined by the target frequency, you can also build one yourself; there are several tutorials online showing you how.

Adapters

If you plan on using other antennas, you'll likely need a supply of adapters to successfully plug them into your SDR's SMA connector. You can accomplish this in one of two ways. First, you can purchase several adapters that have SMA male connectors on one end and a connector compatible with your target antenna on the other. The adapter shown in Figure 12-21, for example, has a male SMA connector on the left and a female coax connector on the right. Simply attach your adapter between your antenna and your SDR.

Figure 12-21: A male-SMA-to-female-coax adapter

The other option is to purchase a kit that contains a number of adapter halves that screw together to build any adapter combination you like. The second option is more expensive than most people need, but it provides the most flexibility.

Either path you choose, you'll most likely need to support adapters for SMA-female to the following male connectors: RP-SMA, MCX, and RCA.

It's best to keep the number of adapters chained together as low as possible, since each adapter in the chain causes some signal loss.

Upconverters

You may have noticed that the HackRF One only works at RF frequencies greater than about 1 MHz. This means you can't just plug it in and start working with lower-frequency signals, like the lower part of the AM broadcast band or certain amateur radio bands. An *upconverter* is a hardware solution to this problem. It behaves just like the downconverters we've built into every one of our radios so far, but instead of shifting the frequency down, it shifts the frequency up. By taking signals that are less than 1 MHz and shifting them into a frequency range above 1 MHz, the upconverter allows the HackRF One to operate on those signals. Like the upconverter shown in Figure 12-22, these devices will typically have SMA inputs and outputs.

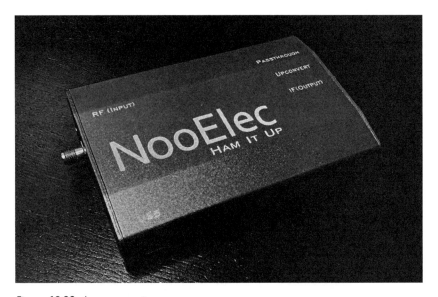

Figure 12-22: An upconverter

Attach your antenna to the upconverter's input and connect your SDR to its output. You 'll also need to apply power, often through a USB cable.

Baluns

A *balun* is a tool that allows a *bal*anced device, like your SDR, to connect efficiently with an *un*balanced device, such as certain antennas. Hence, *bal* + *un* = *balun*. In this context, *unbalanced* simply means the device has a single signal wire plus a ground wire, whereas *balanced* means that both of the wires carry signals relative to each other. Baluns are especially useful when connecting your SDR to homemade antennas, which can sometimes be as simple as a long wire.

Miscellaneous Items

In addition to the major items already mentioned, you should also have a USB cable with a ferrite choke. Also, an SMA-male-to-SMA-female cable can act as an extension cord for your antenna, allowing you to put the antenna farther away from your SDR. This way, you can place your antenna in a more advantageous position, farther away from RF interference, though at the cost of a small reduction in signal strength.

Beyond the basic peripherals that we've covered, the following items could also come in handy in either advanced usage or special scenarios:

Low-noise amplifier (LNA)   This is a hardware gain stage between your antenna and an SDR in receive mode. LNAs provide additional gain for detecting and demodulating weak signals. Most are designed for a limited frequency range, so make sure to check that any LNA you buy covers your target frequencies.

Low-noise block downconverter (LNB)   This device takes signals at frequencies too high for an SDR to process and shifts them down to within the SDR's operating range. These very high frequencies are typically used by satellites or some 5G systems. Think of an LNB as the high-frequency counterpart of the upconverter.

GPS antenna   This is an active antenna specially designed for receiving GPS signals, which are rather weak and hard to pick up with an ordinary antenna. You will need to apply DC power to this antenna, typically through the SMA port. The HackRF One can power this antenna without any extra components, but other SDRs may require a bias tee, which applies DC power to an SMA connection.

Bias tee   This is a wire connecting your SDR to your antenna and will carry the RF power you intend to transmit or receive, but it can also carry a DC voltage to power downstream amplifiers connected to your antenna. If your SDR does not include this DC capability, you can add it with a bias tee, which contains two signal ports and a power connection.

RF attenuator   This reduces the RF power to which your SDR is exposed. High levels of RF power can damage your SDR components. If you think this could be a possibility, you can attach an attenuator between the receive antenna and the SDR. These devices are available in several different attenuation levels, depending on how much reduction is required. Additionally, they will be rated for a maximum power level, so if your receiver is very close to a powerful transmitter, take care to ensure your attenuator can handle the RF power that will be present. Finally, your attenuators will have a specified frequency range over which they're designed to operate.

RF limiter   This caps the RF power that passes through to your SDR but does not otherwise reduce it. You can think of this device as inactive when the RF level is below the limiter's power threshold but restrains the power level from exceeding that threshold if greater power levels are present. In addition to the limiting threshold, these devices will also have a specified frequency range and maximum power over which they can operate.

You probably don't need these specialized components immediately. We've mentioned them so you're aware they exist in case your SDR experiments take you in directions where they come in handy.

Conclusion

This chapter explored different kinds of antennas and how to use them. We also considered other items you might want to gather for your personal SDR toolkit, such as adaptors, upconverters, and baluns. You don't need to run out and buy the whole kit at once, but you should now have an idea of what's available and how it will help you. We've also discussed some strategies for mitigating noise and looked at how your computer's performance can affect your SDR.

Now that you're familiar with antennas and other SDR peripherals, you're ready for the final step in your beginning SDR journey: transmission.

13

TRANSMITTING

Nearly everything we've done so far in this
book has involved *receiving* signals, but what
about *transmitting*? In this chapter we'll build
a flowgraph that will generate the data required
to transmit a broadcast FM signal. We will *not*, however,
hook this flowgraph to a physical SDR. Doing so could
be a violation of the law, as it could essentially jam legitimate FM broadcasts. Instead, we'll execute the flowgraph in software-only mode, running the transmitter's
output directly into an FM receiver flowgraph to verify
it's working. We'll discuss some of the legal and practical ramifications of transmitting in this chapter as well.

Building an FM Modulator

Let's build a transmitter! Start by creating a new flowgraph named *fm_tx.grc* and changing the value of the default samp_rate variable to 8e6. Then create the four variables listed in Table 13-1.

Table 13-1: Transmitter Flowgraph Variables

Block	ID	Value
Variable 1	center_freq	100e6
Variable 2	working_samp_rate	320000
Variable 3	audio_samp_rate	32000
Variable 4	t_width	chan_width/20

Also create a QT GUI Entry block with an ID of freq_tx, a Type of Float, and a Default Value of 98.1e6. Finally, create a second QT GUI Entry with an ID of chan_width, a Type of Float, and a Default Value of 150e3. When you're done, you'll have a lot of variables and an Options block, as shown in Figure 13-1.

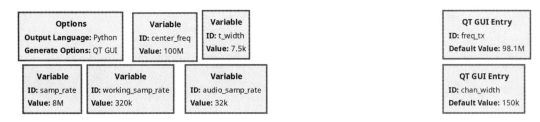

Figure 13-1: The transmitter variables

These variables define all the different frequency and sample rate characteristics we'll need as we build the flowgraph. You've seen most of them before, but we'll do a quick recap of each as you use it.

Setting the Audio Source

Since broadcast FM is a voice medium, we'll use an audio-related source for transmission. If you wish, you can record your own audio, or you can use the *ch_05/HumanEvents_s32k.wav* file from the book's downloads. Add a Wav File Source block, direct it to your chosen file, and set its Repeat option to Yes.

Broadcast FM uses a modulation scheme formally known as *wideband FM*, or *WBFM*. This differentiates it from *narrow-band FM*, or *NBFM*, which uses less bandwidth but has poorer audio quality. The maximum deviation of WBFM, or the amount the frequency can change as the baseband signal modulates the carrier, is 75 kHz. This means we'll need to do the FM modulation at a sample rate at least twice as high as that (remember Nyquist?). To be safe, we'll go a bit further and use a `working_samp_rate` of 320,000. You saw this variable before in the FM receiver project, and you'll see in a moment that it's doing the same thing here that it did there: defining a lower sample rate than the SDR's higher sample rate, represented by `samp_rate`. The next block in the flowgraph will run at this lower sample rate to minimize CPU load.

Modulating the Signal

Next, we'll use a `WBFM Transmit` block to handle the modulation. Connect the block's input to the `Wave File Source` block's output, and set its properties as follows:

Audio Rate: `audio_samp_rate`

Quadrature Rate: `working_samp_rate`

Tau: `75e-6`

Max Deviation: `75e3`

Preemphasis High Corner Freq: `-1.0`

The last three properties should already have the correct values, which are default settings for the broadcast FM standard. The Tau and Preemphasis High Corner Freq properties relate to a little WBFM modulation trick called *preemphasis*. This technique slightly boosts the higher frequencies in the baseband signal, which results in a better signal-to-noise ratio and thus better audio quality. A WBFM receiver will employ the opposite technique, *deemphasis*, to put the baseband signal back where it was. As we've already discussed, the Max Deviation property controls how much the carrier frequency can change due to the modulation. It's equal to 75 kHz, just as we expected.

The Quadrature Rate is the sample rate of the outgoing modulated waveform, while the Audio Rate is the rate for the incoming baseband (the audio signal). This block expects this Quadrature Rate to be an integer multiple of the Audio Rate, which is why we chose our `working_samp_rate` variable to be exactly 10 times greater.

Upconverting the Signal

Now that we have the modulated signal, we need to upconvert it to the correct frequency before sending it out. We'll use a `Multiply` block and a sinusoid carrier. These blocks need to run at a sample rate of 8 Msps. Therefore, before we can implement the upconversion, we need to boost the sample rate of the data yet again. Create a `Rational Resampler` block, connect its input to the `WBFM Transmit` output, and set its properties like so:

Type: `Complex->Complex (Complex Taps)`

Interpolation: `int(samp_rate/working_samp_rate)`

Decimation: `1`

This boosts the sample rate by a factor of 25, bringing the data up to the value of `samp_rate`, which is `8e6`. With that, create a `Multiply` block and pass the Rational Resampler output into one of its inputs. Then connect the other Multiply input to a Signal Source block configured as follows:

Sample Rate: samp_rate

Waveform: Cosine

Frequency: freq_tx - center_freq

Amplitude: 1

The output of this sequence of blocks is the upconverted, modulated signal. Were you to try and transmit this signal, you would hook it up to a Soapy HackRF Sink with a Center Frequency of center_freq and run the flowgraph. The signal would then be physically transmitted at the frequency defined by freq_tx (again, the flowgraph operates with zero-centered frequencies, and the SDR hardware upconverts the zero-centered frequency to the center_freq specified in the SDR sink).

As mentioned before, actually transmitting the signal may be illegal. Rather than run that risk, we'll instead hook up the FM transmitter to an FM receiver flowgraph. First, though, let's see what the transmitter output looks like by passing the output of the Multiply block into a QT GUI Sink with the following properties:

Center Frequency: center_freq

Bandwidth: samp_rate

Show RF Freq: Yes

Your flowgraph should now look like the one shown in Figure 13-2.

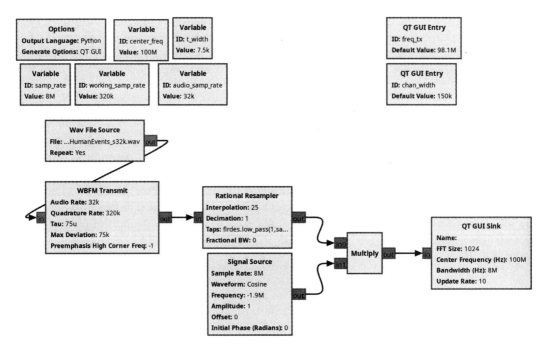

Figure 13-2: The complete transmitter flowgraph

Run the flowgraph, and you'll see an output like Figure 13-3.

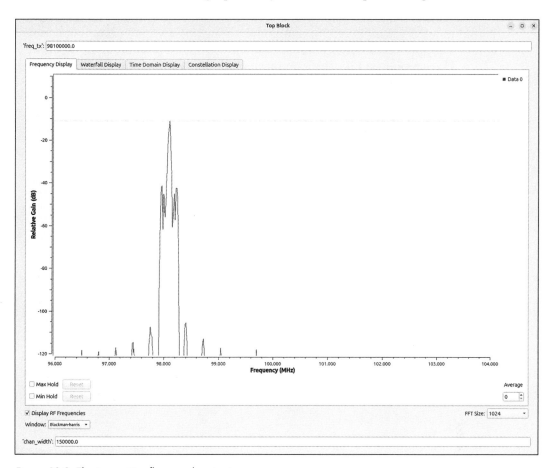

Figure 13-3: The transmitter flowgraph output

Rather than a nice clean spike at the 98.1 MHz transmission frequency, there are several other spikes as well. They're significantly lower and wouldn't cause any trouble, but a basic principle of radio is to ensure that any transmissions are well behaved, consuming the bandwidth necessary and not straying outside into unnecessary frequencies. Therefore, we will show you how to address this problem should you encounter larger and more problematic-sized spikes in the future.

Where did these extraneous spikes come from? If you apply some QT GUI Sinks earlier in the flowgraph, you'll find that the problems start right after the second Rational Resampler. It turns out that interpolation isn't just like decimation in reverse. Simple interpolation will introduce distortion into a signal, and the greater the degree of interpolation, the greater the distortion.

Filtering After Interpolation

To eliminate the distortion spikes, it's best practice to filter a signal after any significant degree of interpolation. Fortunately, the Rational Resampler block has built-in filtering capabilities. Have you ever noticed the block's Taps property? Like the Taps property in the Frequency Xlating FIR Filter block discussed in Chapter 7, it sets the values (called *taps*) needed to implement a finite impulse response (FIR) low-pass filter. There's only one caveat: for the filtering to work properly, you can only interpolate with the block, not decimate. The Decimation property must therefore be set to 1. Fortunately, the ratio between the samp_rate (8e6) and the working_samp_rate (320000) is a simple integer, 25, so decimation isn't necessary.

With this in mind, we'll change the Rational Resampler block to filter the signal after interpolation. Update the Taps field of the Rational Resampler as follows: firdes.low_pass(1, samp_rate, chan_width/2, t_width).

Much like in Chapter 7, we set the Taps property using a Python function called firdes.low_pass(), which calculates the taps for a low-pass filter. We specify that the filter should have a gain of 1 and operate at the samp_rate, with a cutoff frequency of half the chan_width and a transition width of t_width. We use the chan_width variable because its value (150e3) is the expected bandwidth of the WBFM output. Any signal energy at a higher frequency than 75 kHz, or lower than –75 kHz isn't part of the modulated audio and is thus distortion that we want to filter. The transition_width is again computed based on our simple one-tenth rule of thumb.

When you execute the flowgraph, the frequency plot of the transmitter output should now look like Figure 13-4. Looks like the filters removed the distortion.

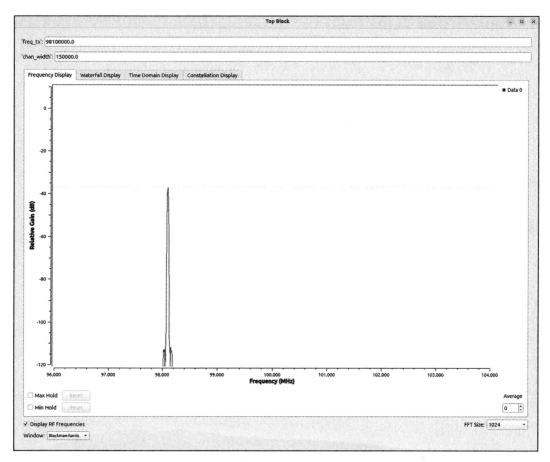

Figure 13-4: The transmitter output with updated `Rational Resampler` *blocks*

Transmission Logistics

Remember the warning that you shouldn't run the transmitter flowgraph through your SDR? Good. It's generally unlawful to transmit on the FM broadcast band without a license. If you look closely into the rules, however, you'll find there are some exemptions. As such, running the transmitter isn't completely outside the realm of possibility, as long as you have a firm understanding of what you're doing and you're very careful to keep the transmitter operating in such a way that it falls into an exempted category. Before we move on to verifying the transmitter is working in software, we should therefore take a moment to discuss the legal and practical issues that arise when using your SDR to transmit.

WARNING *While this section is meant to give you a quick overview of the legal issues around transmitting, it's definitely* not *legal advice. If you run afoul of the Federal Communications Commission (FCC) due to your transmissions, don't point to this book as authoritative legal counsel. It is not.*

Legal Issues

The FCC has governing authority in the United States over how the radio spectrum may be used, and other nations have similar authorities. The FCC divides the RF spectrum into several broad categories. One is the *licensed spectrum*, where companies can purchase or lease the rights to use specific frequencies from the government. This includes cellular telephone companies as well as broadcast radio and broadcast television organizations.

Next, there are the bands that have been set aside for amateur use. Licensed ham radio operators may use these bands, depending on the specific license that they hold. Even licensed hams, however, must still adhere to rules regarding transmission power and other aspects of how they use the spectrum.

Finally, there's the *unlicensed spectrum*, which doesn't require any kind of licensing or cost to operate within. There are still rules governing usage of these unlicensed bands, however. Once again, transmission power is of primary concern. Another key constraint relates to the *duty cycle*, which defines the percentage of time a transmitter is actively transmitting. A transmitter with a duty cycle of 100 percent, for example, is continuously emitting radio signals, while a duty cycle of 50 percent means it's emitting radio energy half the time and emitting no radio energy the other half.

If you consult the relevant FCC regulations for the unlicensed spectrum, you'll often find a table listing pairs of maximum power levels and duty cycles. If you transmit with a lower duty cycle, you'll be allowed to use a higher power level. Accordingly, if you need to transmit more frequently (that is, with a higher duty cycle), you'll need to reduce your power level.

Practical Issues

Say you've spent a lot of time browsing the FCC's website and decided you understand the exemptions well enough to safely and legally operate a private radio station, with very low power, on the broadcast FM band. You've done all that research to ensure the FCC won't hit you with a daily $10,000 pirate radio fine. What next? There are several practical considerations that go into using your SDR as a transmitter, including how to set up your antenna and set the transmit gain.

Antenna Selection and Location

Which antenna would you choose for a low-power FM transmitter? Since you'd probably want an omnidirectional transmission, the ANT500 would work just fine. Extending the antenna all the way would give you roughly one-quarter wavelength and consequently decent gain.

Depending on the nearby landscape, you may need to elevate your antenna a bit to increase its range. On the other hand, since you wouldn't need or want to transmit great distances, you should avoid elevating the antenna any more than is necessary to achieve line-of-sight to your expected receivers. For some bands, you'd need to consider whether to keep your receivers and transmitters outdoors, though the roughly 100 MHz signals in our flowgraph will penetrate most buildings.

Transmit Gain

Our discussion of SDR gain in Chapter 11 focused almost entirely on the receive side of things, but SDR hardware implements gain on the transmit side as well. To recap, the HackRF One transmitter has two hardware gain stages, shown in Figure 13-5.

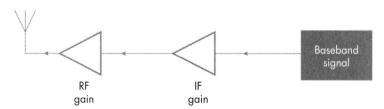

Figure 13-5: The transmit gain stages for the HackRF One

The *RF gain* is an on-off property with a value of either 0 or 14 dB, while the *IF gain* ranges from 0 to 47 dB in 1 dB increments. You can also create a third gain stage in software by inserting a Multiply Const block just before the Soapy HackRF Sink. One problem, though: the Soapy HackRF Sink is only intended to handle a range of values between roughly –1 and +1 (there are complex numbers involved here, and we're being imprecise for the sake of simplicity). It won't produce a legal output outside this range. As long as the peak amplitude of the signal going into the sink doesn't exceed this range, you'll be able to increase the gain in software. If the signal's maximum amplitude is already close to the edge of this range, adding gain in software may cause clipping and distortion.

If you're building a flowgraph that physically transmits, you should start the two transmit gain values (RF and IF) of the Soapy HackRF Sink at 0 dB and see if your receiver is able to pick up a usable signal. You can verify this by viewing a frequency plot of the output of your receiver's Soapy HackRF Source. Look for a spike at the expected frequency. If you don't see a spike, increase the RF gain to 14 dB and check again. Finally, slowly increase the IF gain value until your receiver starts working. You should see the spike at your target frequency increase in size as you increase the gain.

Even though this is just a thought experiment, it illustrates a good rule of thumb: don't use more power than you need to get the job done. Because the HackRF One's maximum output power is relatively low, your transmission range is likely to be fairly limited unless you have a well-positioned antenna with high gain. Even so, follow best practices and try not to turn up your power higher than necessary. Excessive power has several drawbacks. First, it increases the risk of damage to your hardware. Second, you may transmit farther than necessary and possibly interfere with someone else's transmissions. Third, you have a greater likelihood of distortion causing your transmission to spill out of your intended band into adjacent frequencies, again potentially interfering with others.

Testing the FM Transmitter

To say nothing of the legal issues, we're going to assume that you don't have two SDRs: one to run the FM transmitter and another to test it with an FM receiver. Instead, we'll finish off the transmitter project (and this book!) by adding a software FM receiver to the flowgraph to verify that the transmitter portion is working correctly. In fact, even if you had the necessary hardware to run both the transmitter and the receiver, it's still a good idea to simulate both ends of the system in software whenever possible so you can find and eliminate errors before getting the hardware involved.

To understand how the simulated transmitter-receiver system will work, first imagine building hardware versions of each radio. When you operate them both, the entire system would look something like Figure 13-6.

Figure 13-6: A high-level view of a hardware-based WBFM system

On the transmit side, the input signal goes through the transmit flowgraph for processing, and the result is sent to an SDR for transmission via an antenna. On the receive side, another antenna picks up a signal and passes it through an SDR into the receive flowgraph, which processes the captured RF data to produce an output signal that ideally matches the original input. Conceptually, what we're going to do now is cut out the SDR hardware and connect the transmitter directly to the receiver, making this connection entirely inside the flowgraph. At a high level, the resulting system will look like Figure 13-7.

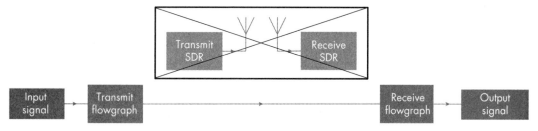

Figure 13-7: A high-level view of a simulated WBFM system

Our goal, then, is to add the software components of our previous hardware FM receiver flowgraph (see Chapter 9) to the current transmitter flowgraph.

Recovering the Signal

The first step in a receiver is to tune to the incoming signal. Fortunately, most of the variables you'll need have been defined already on the transmitter side. You'll just need a new GUI control to control the receiver's tuner. Add a QT GUI Entry and configure it as follows:

ID: freq_rx

Type: Float

Default Value: 98.1e6

Previously, our FM receiver's Soapy HackRF Source fed into a Frequency Xlating FIR Filter block, so we'll use one of those. Add the block to the flowgraph, connect its input to the Multiply block's output, and adjust the following settings:

Decimation: int(samp_rate/working_samp_rate)

Taps: firdes.low_pass(1, samp_rate, chan_width, t_width)

Center Frequency: freq_rx - center_freq

Sample Rate: samp_rate

This implements tuning by downshifting, filtering, and decimating the received signal. From there, we can demodulate, filter, and resample it to extract the original audio signal. For the demodulation, connect the Frequency Xlating FIR Filter output to the input of a WBFM Receive and set the latter as follows:

Quadrature Rate: working_samp_rate

Audio Decimation: int(working_samp_rate/audio_samp_rate)

The output from the WBFM Receive block will now contain demodulated audio, with its sample rate reduced by a factor of 10, from 320 ksps down to a rate of 32 ksps. In order to hear the output, connect the WBFM Receive to an Audio Sink with a sample rate of 32 kHz. The flowgraph should now look like Figure 13-8.

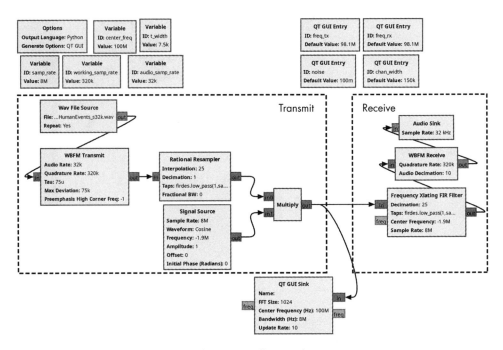

Figure 13-8: The combined transmitter-plus-receiver flowgraph

You can see the transmit portion on the left half and the receive portion on the right half of the figure. Be sure to compare the receive portion to your previous FM receiver flowgraph to verify that they're equivalent.

Running the Flowgraph

Run the flowgraph. Since the default QT GUI Entry values for freq_tx and freq_rx are the same, the receiver is tuned to the same frequency as the transmitter. You should therefore hear some audio, though you may have to increase either the volume value in the flowgraph or the volume of your computer's speakers. If the audio is clear, you've got a working transmitter.

While you're here, though, try retuning your receiver to a different channel. Set freq_rx to any value that seems interesting, but leave freq_tx alone. This means that your transmitter is still sending out a signal at 98 MHz, but your receiver will be tuned to a different frequency. For most of the other frequencies you might randomly choose, you should hear either nothing or a bit of static. But try something specific: set freq_rx to 98.4M, as in Figure 13-9.

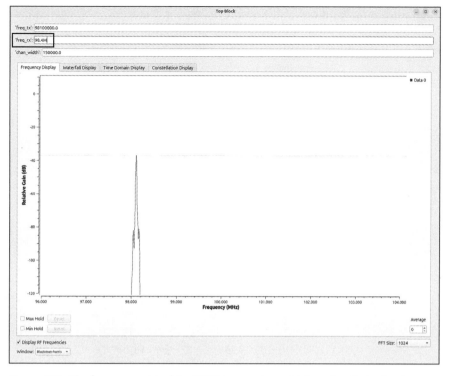

Figure 13-9: The frequency plot of the WBFM transmitter with intentionally bad tuning

Hmmm . . . why are you still hearing audio when you're tuned to a different frequency? Even though it's somewhat distorted, the sound should still be recognizable. Because of this, you might think that your flowgraph is flawed in some way. Strangely, though, this effect is happening more for the opposite reason: your flowgraph is insufficiently flawed!

Two things are working together to cause this effect. First, the WBFM modulation process causes a significant amount of *harmonic distortion*. This means that in addition to the modulated signal at the intended frequency, there will also be much smaller copies of that signal, known as *harmonics*, on each side of the target frequency. In fact, there won't be just one copy on each side, but an infinite series of copies at regularly spaced frequency intervals that tend to diminish in strength as you get farther from the target frequency, as illustrated in Figure 13-10.

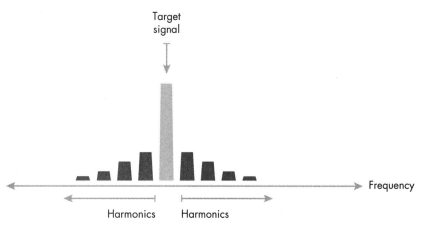

Figure 13-10: A frequency-domain representation of harmonics

Typically these harmonics wouldn't be an issue. They're already relatively small, and they're diminished further by the filter in our second Rational Resampler block. As small as these harmonics are, however, the WBFM Receive block seems to detect them. That's where the second issue comes to light. Because the transmitter-receiver flowgraph is operating in an artificial environment almost completely free of noise, the signal-to-noise ratio of these harmonics is still significant. Normally the receiver would be picking up all sorts of noise from the environment that would drown the harmonics out, but the only noise in an all-software system is computation and quantization noise. Because all the math is being done with 32-bit floating-point values, this mathematical noise is almost nonexistent. How much noise is there? Look at the QT GUI Sink on the output of your transmitter, and scroll down with your mouse wheel until you see something like Figure 13-11.

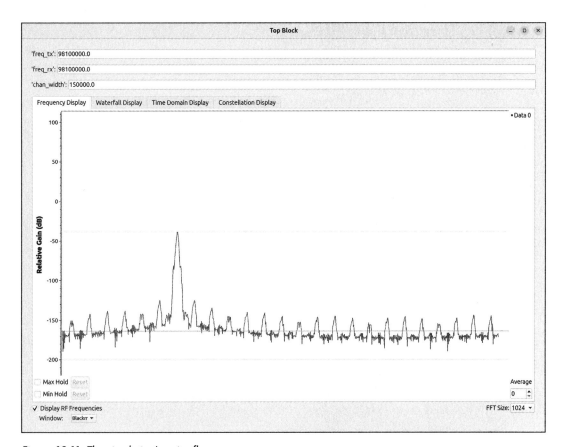

Figure 13-11: The simulation's noise floor

When you scroll down, you'll begin to see the evenly spaced peaks from the extra harmonics. You should also see the noise floor all the way down near –200 dB. This is an incredibly low value, one that doesn't occur naturally on Earth. As such, it doesn't take much of a signal, or a harmonic, to be greater than –200 dB.

Modeling Noise

We've discussed how the lack of noise is causing the flowgraph to behave oddly. What can we do about it? How about trying to make the flowgraph a little bit more like reality by adding some artificial noise between the transmit and receive portions? This noise should be set to a level similar to that encountered in a real SDR system. When you operated your previous hardware-based FM receiver, you should have seen a noise floor at roughly −80 dB, so we'll try for that.

Break the connection between the Multiply and Frequency Xlating FIR Filter blocks. Then create a Noise Source block with the following settings:

Noise Type: Uniform

Amplitude: 0.01

The amount of noise this block generates is determined by the Amplitude property, although we unfortunately can't set this in decibels. We've given you the value of 0.01, which corresponds to the −80 dB noise level we're looking for. If you want a different level, try setting the Amplitude property dynamically with a QT GUI Entry and iteratively input values until you get the noise floor you want. The function of the Noise Type property is beyond the scope of this book.

Next, create an Add block with two inputs. Connect the first input to the Noise Source block's output and the second input to the Multiply block's output. Then pass the Add output into the Frequency Xlating FIR Filter block input. In effect, this combines the noise with the intended transmission before the resulting RF data is picked up by the receiver.

Finally, move the QT GUI Sink connection to the output of the Add block so that you can see the signal plus the noise rather than just the pristine signal. When you're done, you'll have something like Figure 13-12.

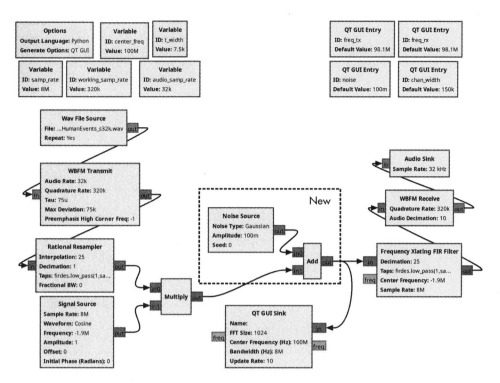

Figure 13-12: The final WBFM flowgraph with noise modeling

Execute the flowgraph and make sure the `freq_rx` value is set back to
98.1e6. You should now see the noise on either side of the signal has increased
to more normal levels, as in Figure 13-13. Looks a lot more like the FM
receiver flowgraph when you ran it with hardware, doesn't it?

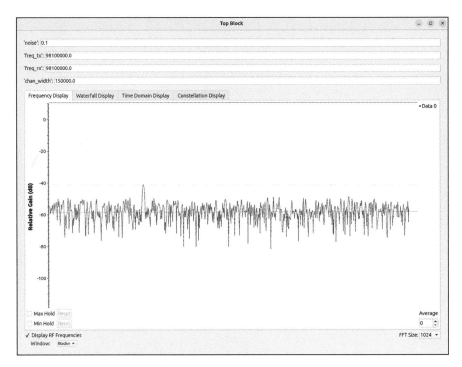

Figure 13-13: The final WBFM frequency plot with noise

Despite the noise, you should still hear the audio playing. This means the noise hasn't impacted the proper function of your transmitter or receiver. When you tune to a different frequency, however, by changing freq_rx to 98.4M (or any other empty frequency), you should no longer hear the distorted audio, just the sound of static. The signal-to-noise ratio of the extra harmonics is too low for the harmonics to be distinguishable from the noise.

The key lesson here is that you should always add noise to your simulated flowgraphs to ensure that your radio will work as expected when you plug in actual SDR hardware.

Conclusion

In this chapter, you built your first transmitter and learned how to legally test it in a somewhat-realistic simulation environment. Furthermore, you saw how you would implement your transmitter with SDR hardware and how to adjust the gain. You also got some non–legally binding advice on what to consider when operating a hardware radio to ensure compliance with the rules.

You've made it to the end of your introductory SDR journey! You should now be able to build basic transmitters and receivers with GNU Radio Companion. You've been introduced to the basic theory underpinning these radios. And finally, you've seen how to use SDR devices and what kind of hardware accessories might be useful.

INDEX

Practical SDR is set in New Baskerville, Futura, Dogma, and TheSansMono Condensed.